Green Energy and Technology

Climate change, environmental impact and the limited natural resources urge scientific research and novel technical solutions. The monograph series Green Energy and Technology serves as a publishing platform for scientific and technological approaches to “green”—i.e. environmentally friendly and sustainable—technologies. While a focus lies on energy and power supply, it also covers “green” solutions in industrial engineering and engineering design. Green Energy and Technology addresses researchers, advanced students, technical consultants as well as decision makers in industries and politics. Hence, the level of presentation spans from instructional to highly technical.

Indexed in Scopus.

Indexed in Ei Compendex.

Allana Katiussya Silva Pereira ·
Ananias Francisco Dias Júnior
Editors

Agro-industrial Waste for New Pharmaceuticals

Sustainable Sources of Bioactive Compounds

Springer

Editors
Allana Katiussya Silva Pereira
Department of Forest Sciences
"Luiz de Queiroz" College of Agriculture
(Esalq)
University of São Paulo (USP)
Piracicaba, São Paulo, Brazil

Ananias Francisco Dias Júnior
Department of Forestry and Wood Sciences
Federal University of Espírito Santo (Ufes)
Jerônimo Monteiro, Espírito Santo, Brazil

ISSN 1865-3529 ISSN 1865-3537 (electronic)
Green Energy and Technology
ISBN 978-3-031-88536-5 ISBN 978-3-031-88534-1 (eBook)
https://doi.org/10.1007/978-3-031-88534-1

This Springer imprint is published by the registered company Springer Nature Switzerland AG
The registered company address is: Gewerbestrasse 11, 6330 Cham, Switzerland

Foreword

Agro-industrial waste is part of the daily reality of various productive sectors, from highly industrialized countries to developing nations. These residues, once considered emblematic environmental problems, are now regarded as indispensable opportunities for the development of new products. Such products, in addition to fostering more sustainable solutions from economic and social perspectives, can help mitigate the environmental impacts caused by the unregulated accumulation of this type of waste while also promoting positive outcomes within the circular economy. One of the applications that has gained prominence due to its potential is the use of these residues in products related to human health, whether through direct or indirect applications, aiming to generate overall improvements that positively impact people's lives. Addressing the impacts of these residues on the development and creation of new products is an urgent matter, and this book explores these issues from both multidisciplinary and interdisciplinary perspectives.

This book emerges from the concerns of the organizers and authors who, by dedicating their studies and research to exploring new possibilities for the use of agro-industrial waste, recognized an opportunity to examine this issue in greater detail. As such, this work brings together various reflections on the current scenario and the upcoming challenges surrounding the development of products that utilize waste as raw material. Moreover, the authors enable readers to understand, through discussions of case studies currently under investigation, that different types of waste can follow distinct reuse pathways. In other words, with proper waste characterization and a comprehensive understanding of its application aspects, it is possible to devise efficient strategies that maximize the benefits of transforming waste into valuable products.

Through the topics covered in this book, readers will have the opportunity to broaden their perspectives and engage in more comprehensive discussions about the challenges posed by materials discarded by the agro-industrial sector. Additionally, they will be able to explore real possibilities for applying such waste in the development of new products in the health and environmental sectors. The discussions presented herein encourage readers to draw their own conclusions about innovative uses that add value to agro-industrial waste, while addressing future perspectives, challenges, and opportunities that make this material so promising. Supported not only by their personal insights and research experiences but also by a robust review of relevant literature, the authors establish critical questions that enrich the scientific discourse. This combination of perspectives makes the book particularly engaging and valuable from a scientific point of view.

To ensure that readers have the opportunity to explore, question, and discover for themselves the challenges associated with the use of agro-industrial waste in the production of new medicines and its potential contributions to sustainable development, I conclude my words by wishing you an enriching reading experience. May the discussions presented in this book inspire researchers from various fields and regions across the globe. May more waste be repurposed to generate greater benefits for society—whether environmental, economic, or social. Strategies that contribute to sustainable development are of paramount importance, particularly in today's world. It is my hope that this book offers valuable insights, encouraging you, dear reader, to reflect on and implement these ideas within your own reality.

Elias Costa de Souza
Federal University of Southern
and Southeastern Pará
Marabá, Brazil
eliascosta@unifesspa.edu.br

Contents

Agro-industrial Waste as a Potential Raw Material for Multiple Products and Promotion of a Circular Economy

Paulo Renato Souza de Oliveira, Kamilla Crysllayne Alves da Silva, Gabriela Aguiar Amorim, Saly Takeshita Yamaguti, Daniel Saloni, Allana Katiussya Silva Pereira, and Ananias Francisco Dias Júnior

Abstract The amount of waste produced by the agricultural and agro-industrial sectors is a global concern. What constitutes agro-industrial wastes? What are their disposal routes, and how can industries reuse them? While disposal methods may vary, these wastes are often improperly managed, causing significant environmental harm. However, there is considerable potential to repurpose this waste, transforming it into new products, including pharmaceuticals, thus reducing waste. This chapter presents a comprehensive overview of the different types of waste generated by the agricultural and agro-industrial sectors. It highlights the variety of raw materials available and how they can be repurposed to develop new medicines. We

P. R. S. de Oliveira (✉) · A. F. Dias Júnior
Department of Forestry and Wood Sciences, Universidade Federal do Espírito Santo (Ufes), 316 Av. Governador Lindemberg, Jerônimo Monteiro, ES 29550-000, Brazil
e-mail: pr.oliveira0@gmail.com

A. F. Dias Júnior
e-mail: ananias.dias@ufes.br

P. R. S. de Oliveira · K. C. A. da Silva · G. A. Amorim · A. K. S. Pereira
Department of Forest Sciences, "Luiz de Queiroz" College of Agriculture (Esalq), Universidade de São Paulo (USP), 11 Av. Pádua Dias, Piracicaba, SP 13418-900, Brazil
e-mail: kamialves97@gmail.com

G. A. Amorim
e-mail: gabriela.a.amorim@hotmail.com

A. K. S. Pereira
e-mail: allana.florestal@gmail.com

S. T. Yamaguti
Universidade Federal da Amazônia (UFRA), Campus Paragominas, Rodovia PA 256, Nova Conquista, 68627-451, Paragominas, PA, Brazil
e-mail: saly.takeshita@ufra.edu.br

D. Saloni
Department of Forest Biomaterials, College of Natural Resources, North Carolina State University (NCSU), 411 Dan Allen Drive, Raleigh, NC 27695, USA
e-mail: desaloni@ncsu.edu

A. K. Silva Pereira and A. F. Dias Júnior (eds.), *Agro-industrial Waste for New Pharmaceuticals*, Green Energy and Technology,
https://doi.org/10.1007/978-3-031-88534-1_1

explore the potential of agro-industrial waste as a valuable resource in the transition to a circular economy. These sectors produce substantial waste, which, if not appropriately managed, can have negative environmental, public health, and socioeconomic impacts. Traditional linear economy models have treated these materials as residues; however, recent research shows their potential for transforming into high-value products like bioactive compounds, pharmaceuticals, biofuels, and bioplastics. The chapter discusses various technological approaches for processing and valorizing these residues, including pyrolysis, gasification, and biotechnological methods. These processes not only minimize waste accumulation but also facilitate the extraction of valuable bioactive compounds, supporting a more sustainable and economically viable industry. Aligned with the United Nations Sustainable Development Goals (SDGs), this chapter emphasizes integrating waste valorization strategies into agro-industrial practices to promote environmental responsibility and resource efficiency. It also highlights the importance of policy support, technological innovation, and collaboration between industries and governments to implement effective, sustainable waste management solutions.

Keywords Sustainable development · Value-added products · Waste valorization · Bioactive compounds · Decarbonization

1 Paradigm Shift in Global Production Models: Limitations and the Need for Change

Manufacturers' current consumption and production patterns have proven predominantly unsustainable, endangering environmental balance and generating societal consequences [137]. The traditional approach, known as the linear economy, brings significant negative impacts such as ecosystem degradation, excessive waste generation, and increased carbon emissions. These effects not only pose risks to the environment but also threaten human health and global well-being [103]. The main limitation of the linear economy is its inability to capture value from products after their use, leading to inefficient resource exploitation and, consequently, higher production costs. In this model, products are discarded once they lose functionality or utility, disregarding possibilities for reuse, recycling, or recovering materials that still hold value [2]. This limitation reflects an outdated mindset that assumes natural resources are infinite and easily accessible, ignoring the growing scarcity of raw materials and the urgency to reduce environmental impacts.

The linear system operates on the take-make-use-dispose model. Companies extract raw materials, transform them into products, and sell them to consumers. These products are discarded at the end of the lifecycle, often without reuse or material recovery. This model generates vast amounts of waste and contributes to the depletion of natural resources and worsening the environmental crisis [135]. In this context, agro-industrial waste—including residues from agriculture, forestry, and livestock activities—plays a critical role. These wastes, often underutilized or discarded, are generated at various stages of the production chain, including handling, processing, production, storage, distribution, and consumption [4, 115].

The transition from a linear economy to a more sustainable model, such as the circular economy, is directly aligned with the Sustainable Development Goals (SDGs), which provide a global framework for sustainable development. A key principle of the SDGs related to circular economy concepts is the responsible use of natural resources. Considering this, new approaches have been developed to assist industries and policymakers in adopting more circular practices, promoting the reuse, recycling, and recovery of materials to reduce waste and minimize environmental impacts.

The circular economy emerges as a response to the challenges posed by the linear economy, offering innovative solutions to maximize product value throughout their lifecycles. The circular economy (CE) is a system based on reducing, reusing, and recycling to create a closed-loop cycle of production and consumption. This model aims to minimize the use of new resources while reducing waste generation and greenhouse gas emissions [93, 117]. Such a shift is essential for ensuring environmental sustainability and critical for achieving global targets established by the SDGs, promoting a more resilient future for upcoming generations. The improper management of agro-industrial waste can result in severe environmental consequences, including greenhouse gas (GHG) emissions that aggravate global warming [21]. When not treated sustainably, these byproducts significantly threaten the environmental balance. Due to limited air circulation and moisture in waste piles, anaerobic conditions may develop, promoting CH_4 formation. Additionally, nitrification and denitrification processes generate N_2O. These GHG emissions lead to carbon and nitrogen losses, intensifying global warming potential [180]. Thus, transforming waste into renewable resources through processes such as bioenergy production and bioproduct generation [3, 86], becomes an urgent necessity for mitigating environmental impacts and promoting sustainability and resource efficiency.

Applying circular economy principles in agriculture, forestry, and livestock industries aims to transform agro-industrial waste into valuable resources, closing the production cycle and reducing waste generation. Many of these byproducts hold significant reuse potential, as they are rich sources of biopolymers such as cellulose, starch, pectins, and proteins. Furthermore, they contain bioactive compounds like polyphenols, flavonoids, vitamins, tannins, natural pigments, and essential oils that can be utilized in various industries, including food, pharmaceuticals, and cosmetics [65]. Numerous technological and biotechnological processes have been implemented to convert agro-industrial waste into value-added materials. Notable among these are pyrolysis, gasification, and carbonization, which transform plant residues into biofuels, charcoal, and other valuable products, maximizing energy and resource use efficiency [133]. Utilizing agro-industrial waste as raw materials across industries not only reduces environmental impact but also creates economic opportunities, fostering innovation and sustainable development.

This focus on circular economy within agro-industry reveals tremendous potential for waste reduction while enhancing resource utilization. With proper management and continuous technological advancements for byproduct reuse, the circular economy can contribute to creating more efficient production chains, generating value at every stage—from cultivation to the final waste destination.

2 Agro-industrial Waste: Concepts and Valorization into Multiple Products

Various human activities rely on products derived from agriculture, forestry, and livestock, whether for food production or the creation of other consumer goods. It is thus possible to classify numerous products based on the nature of the material that generated them or the processes they underwent. Regardless of classification, agricultural, forestry, and livestock production inevitably generate co-products or waste. These materials may be inedible or undesirable as final products arise from operations such as planting, handling, harvesting, beneficiation, processing, or even product consumption. Ogbu and Okey [121] define waste as unwanted items during or after agricultural or agro-industrial operations. In this context, the vast diversity and specificities of the agricultural, forestry, and livestock sectors result in substantial and heterogeneous waste production.

2.1 Classification and Sources of Waste

Agro-industrial waste can be broadly categorized into two main spheres: on-farm and off-farm (Fig. 1).

In Fig. 1, two subcategories can be identified: field waste and processing waste. Seeds, leaves, stalks, bunches, and other agricultural products left in the field after harvesting are classified as field waste, while process waste includes husks, roots, and seeds generated after the processing of raw products [160]. Additionally, forestry materials such as stumps, branches, bark, roots, fruits, and other non-timber materials [110] can also be included in this category. On the other hand, off-farm waste is generated in food processing industries and includes husks, oil, bagasse, soil, and molasses. Forestry-based industries, such as pulp and paper production, also generate waste in sludge, wood fibers, and recycled paper [114], with black liquor being a notable byproduct of the Kraft pulping method. Primary and secondary manufacturing of hardwood and softwood also contribute to forest residues, including bark, sawdust, cuttings, and wood chips [111]. It is important to note that residual biomass diversity is not uniform across all regions, specific production chains influence it over time. Countries with extensive arable land and climates favorable to agriculture tend to have greater availability of waste biomass. In this context, India, the United States, China,

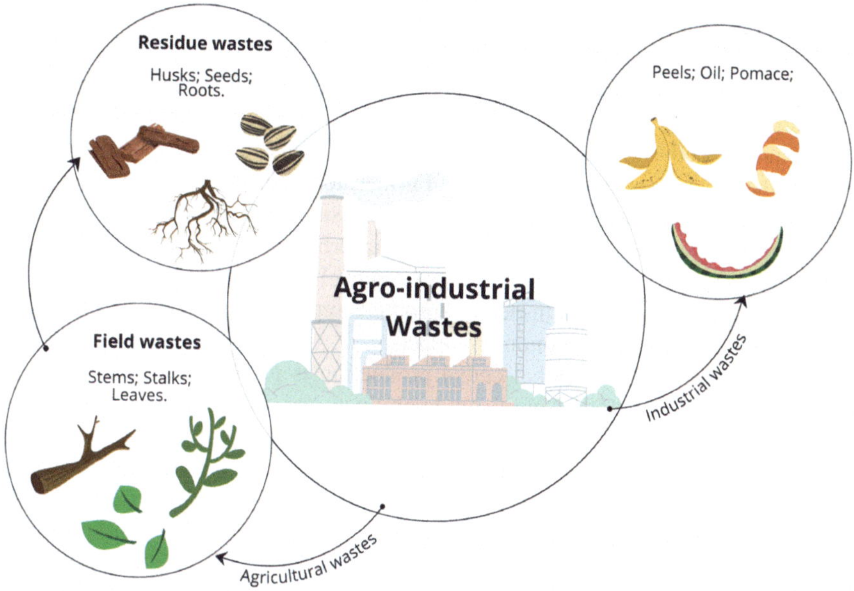

Fig. 1 Classification of agro-industrial waste based on source and processing level. *Source* Authors (2024)

and Russia stand out, collectively contributing over 577 million hectares of cultivated land globally. Brazil ranks fifth with 63 million hectares, followed by Indonesia, Nigeria, Argentina, Canada, and Ukraine [50]. These nations have an advantage in terms of raw material availability, particularly regarding on-farm waste. It is worth noting, however, that countries that primarily export their products as commodities (e.g., China, Brazil, and Indonesia) or have lower levels of industrialization will have a smaller share of off-farm waste.

2.1.1 Agriculture Wastes

Globally, food production has increased due to intensive farming practices and high demand. This growth has also elevated waste generation. Estimates indicate that over 5.2 billion tons of waste were generated from agricultural crops in 2020 [153], based on the Food and Agriculture Organization (FAO) report of 18.7 billion tons of agricultural production. By 2022, this waste increased further, as agricultural production rose by approximately 441 million tons between 2020 and 2022 [57, 58]. In India, the world's most populous nation with over 1.4 billion inhabitants [175], agricultural waste generation is estimated to be between 350 and 990 million metric tons annually [91]. Major contributors include rice (*Oryza sativa* L.), wheat (*Triticum aestivum* L.), sugarcane (*Saccharum officinarum* L.), and cotton (*Gossypium hirsutum* L.).

However, only 25–30% of this waste is reused; the remainder is either burned or improperly discarded [108].

It is also estimated that approximately 25% of the global agricultural biomass is directly burned [177], contributing significantly to agriculture being the second-largest human activity in terms of greenhouse gas emissions (19.9%), surpassed only by the energy sector (68.1%) [95]. Furthermore, agricultural expansion into high-carbon ecosystems, such as tropical rainforests and other natural habitats in Latin America, Southeast Asia, and Africa, exacerbates the challenge of building a resilient, sustainable, and low-carbon economy [41, 79, 85]. Reusing agricultural waste to replace petroleum-derived products represents a dual opportunity for positive environmental and economic impacts.

According to the FAO, global cereal production is projected to reach 2,847.4 million tons in 2023/24 [50]. Significant byproducts and waste are generated alongside this production during harvesting and processing. Cereal processing contributes 12.9% to global food waste [63]. Byproducts like bran, germ, and grains are rich in valuable compounds such as dietary fibers, proteins, essential fatty acids, and antioxidants, yet are often underutilized as animal feed or discarded [51, 163]. Promising prospects exist for converting this waste into higher-value products. Song et al. [164] explored rice husk as feedstock for biorefineries, demonstrating high efficiency in co-producing bioethanol, lactic acid, and D-psicose[1] through chemical pretreatments and enzymatic hydrolysis. The annual production of approximately 150 million tons of rice husk also finds applications in construction materials, pharmaceuticals, and nanotechnology [88].

Barley malt (*Hordeum vulgare* L.) is another cereal widely used in the food and beverage industries, particularly baked goods, whiskey, and beer. The brewing industry generates malt bagasse or spent grain, which accounts for approximately 85% of its raw materials. The composition of this residue varies based on grain variety and harvest season [178]. With global beer production reaching 1.88 billion hectoliters in 2023 [19], valorizing spent grain through multi-product development could enhance sustainability and create eco-friendly inputs and bioproducts. Buller et al. [25] highlighted spent grain as a clean energy source for low-carbon brewing. Pretreated biomass with ultrasound showed excellent results in anaerobic digestion processes for biogas production. Other studies explored its incorporation into biodegradable films as alternatives to plastic packaging [157], its use as a low-cost biosorbent for textile dye removal [53], and its conversion into biochar for adsorbing contaminants in wastewater [158].

The latest *Citrus: World Markets and Trade* report by the U.S. Department of Agriculture's Foreign Agricultural Service [170, 171] shows that global citrus fruit production exceeded 102.6 million metric tons. Key varieties include orange (*Citrus sinensis* (L.) Osbeck), tangerines/mandarins (*Citrus reticulata* Blanco), grapefruits (*Citrus paradisi* Macfad.), and lemons/limes (*Citrus limon* (L.) Osbeck/*Citrus aurantiifolia* (Christm.) Swingle), with respective productions of 47.4 million, 38.2 million,

[1] It is a monosaccharide, an epimer of fructose, uncommon in nature. It is considered a low-calorie sugar and is commonly used as a sweetener in foods and beverages.

6.9 million, and 10.1 million tons for 2023/24. These fruits' residues hold great potential due to their abundance and composition, which are rich in high-value compounds. Pectin, for example, is a prebiotic polysaccharide with health benefits, offering gelling, thickening, water retention, and encapsulation properties [152]. This makes citrus pectin suitable for the food, pharmaceutical, and personal care industries [161]. Additionally, citrus byproducts such as pressed pulp, peels, leaves, seeds, and secondary juice are rich in phytochemicals and other valuable compounds, including polyphenols, palmitic and linolenic acids, proteins, carotenoids, monoterpenes, and vitamins C and B complex [104]. Wood biomass from unproductive citrus trees also presents opportunities. For example, *C. sinensis* residual wood biomass underwent slow pyrolysis to produce bio-oil. Following bi-distillation, the bio-oil extract demonstrated antiviral and antibacterial properties [172–174].

In line with this scenario, tropical fruits also represent a significant portion of global food commodities. According to FAO data, the main commodities in the market include mango (*Mangifera indica* L.), mangosteen (*Garcinia mangostana* L.), guava (*Psidium guajava* L.), pineapple (*Ananas comosus* (L.) Merr.), avocado (*Persea americana* Mill.), and papaya (*Carica papaya* L.). Together, these fruits accounted for over 8.7 million tons produced in 2023 [50]. Tropical fruits play a vital role as sources of vitamins and minerals in human diets. They are widely consumed, with their residual biomass also being rich in cellulose and hemicellulose polymers and lignin macromolecules. These can facilitate the creation of higher-value products derived from their bioactive compounds [26]. Ding et al. [42] summarized that physical pre-treatment processes or anaerobic digestion, followed by enzymatic hydrolysis and fermentation, can yield numerous value-added products, including energy inputs, bioactive compounds, polyphenols, nanomaterials, and biofertilizers (Fig. 2).

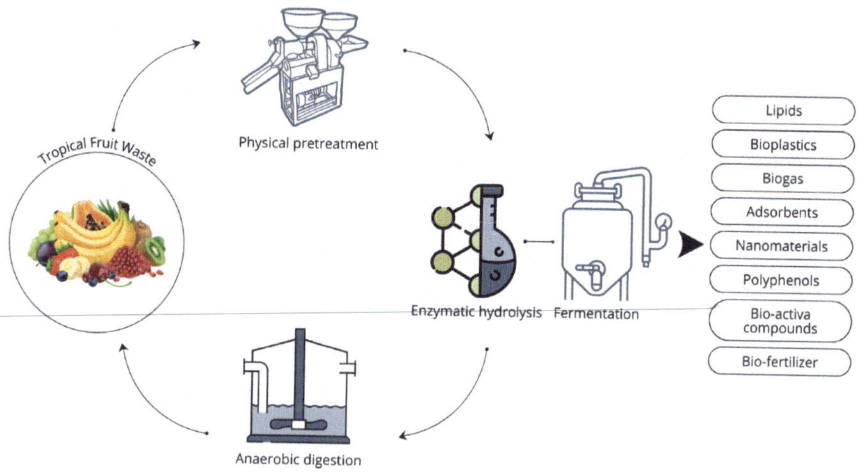

Fig. 2 Diagram of the utilization of agroindustrial waste from fruit processing to produce multiple products. *Source* Authors (2024)

Additionally, the same authors indicate that microorganisms capable of degrading lignin can be used in bioprocesses for the production of bioproducts, or their enzymes can be employed in the hydrolysis of lignocellulosic substrates, facilitating their conversion and utilization by different microorganisms. Recent studies have also demonstrated the potential of tropical fruit waste for producing glucose and ethanol through pre-treatment processes, enzymatic hydrolysis, and fermentation [37]. The high phytochemical content in components of seeds, flowers, leaves, and peels is also promising in the treatment of obesity, offering a natural alternative to synthetic drugs [15]. In general, the utilization of tropical fruit by-products, rich in bioactive compounds, is increasingly generating interest in understanding how these wastes can be converted into valuable resources for the prevention or treatment of health issues, such as cardiovascular diseases, diabetes, neurological conditions, and cancer, particularly related to oxidative stress [26]. New discoveries and the development of effective processing techniques must occur to maximize the utilization of these wastes.

Specifically in the Pan-Amazon region,[2] residual biomass from plants of socio-biodiversity stands out, with their food consumption and use in traditional medicine potentially driving more in-depth research. A review of studies on Amazonian plants reveals that inedible parts of various plant species are rich in bioactive compounds with great potential for medical, pharmaceutical, and even food industry applications [98]. One example is açaí (*Euterpe oleracea* Mart. and *Euterpe precatoria* Mart.), a palm tree that provides fruit and hearts of palm. Traditionally present in the diets of native populations, açaí has gained global prominence as a "superfood," increasing the demand for its fruits in the production of higher-value pharmaceutical and food products, particularly since the 2000s [23]. As a result, waste generation has also increased, considering that about 80–85% of the fruit consists of inedible parts [24, 127]. This volume represents a significant fraction of the total production, which reached approximately 1.7 million tons in 2023 [76].

Consequently, several studies have explored the use of this residual biomass in various areas, such as the production of densified fuels (briquettes) [38], bio-oil [122], activated charcoal [39, 130], soil conditioners [147], nanoproducts [34, 69], and even for the production of active ingredients for anti-tumor drugs [34]. It is also worth highlighting initiatives aimed at producing kraft pulp and paper with fibers from açaí waste as an alternative for value addition on an industrial scale while simultaneously engaging the local community [123]. Overall, the Amazon Rainforest, with over 14,000 species of native plants [30] and various agricultural systems, stands out as an essential territory for the utilization of residual biomasses.

---

[2] The Pan-Amazon region encompasses parts of nine different countries: Brazil, Bolivia, Colombia, Ecuador, Peru, Venezuela, Suriname, Guyana, and French Guiana.

2.1.2 Forest Wastes

Biomes and forested areas cover 31% of the Earth's land surface, totaling 4.06 billion hectares. Of this area, planted forests cover 294 million hectares, corresponding to 7% of the global forest area [56]. In this context, the forestry sector, including management and planting, generates large amounts of biomass waste that can be utilized for various purposes (Fig. 3). In general, we can categorize these wastes as field wastes (from silvicultural practices and the harvest of wood and other forest products), mechanical processing wastes (obtained from sawing and refining wood into lumber or engineered wood products), and industrial wastes, generated by chemical and thermochemical conversion routes (present in the paper and pulp production chains, as well as charcoal production).

Globally, the production of biomass derived from wood is estimated at around 4.6 Gt annually, with 60% of this total being used for energy generation, 20% for industrial use as round wood, and the remaining 20% representing primary losses that remain in the field for decomposition. It is estimated that 80% of the mass of forest trees is discarded as waste, with only 20% being converted into kiln-dried sawn products [54]. Furthermore, the generation and recovery of waste, both in the field and in industry, depend on factors such as the tree species used and geographic conditions. For every cubic meter of wood processed, another cubic meter of waste remains in the forest. In the primary processing stage, which occurs in industries, the waste includes branch pruning and bark removal (about 12% of the material reaches sawmills), additional blocks and chips (about 34%), and sawdust (about 12%). After kiln drying, chips (about 6%) and more sawdust/pruning (about 2%) increase the waste volume [55, 77]. According to estimates from the production of

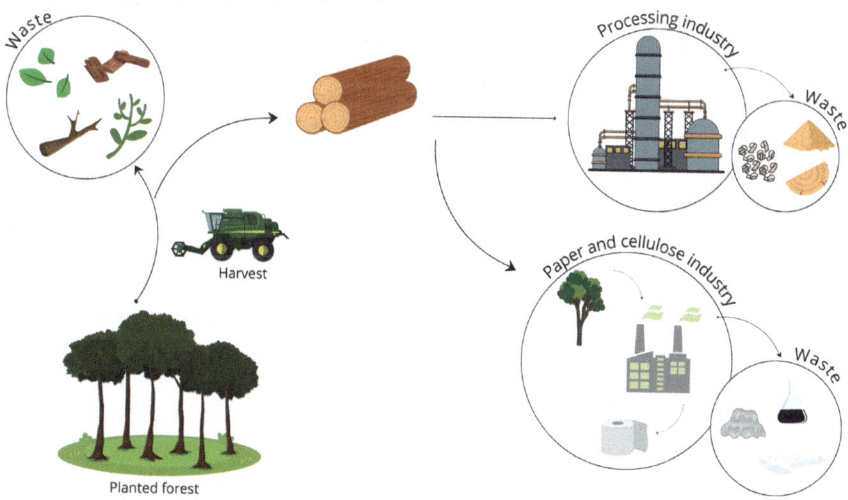

Fig. 3 Forest wastes generated based on the source and the production process of forest-based products. *Source* Authors (2024)

waste in productive forests across 21 countries [54], Tripathi et al. [169] indicate that 700 Mt year^{-1} of forest residues could be used in the production of various products and energy generation.

According to recent data, the European Union (EU) area of forests available for timber exploitation is approximately 160 million hectares [49]. This represents about 39% of the total area of Europe. In 2021, the wood stocks in EU forests amounted to approximately 29 billion m^3. Germany, Sweden, and France were the leading contributors from 2000 to 2021. The increase in wood stocks is primarily due to the expansion of forest areas through reforestation and natural regeneration [49]. Intensifying the use of this woody biomass by optimizing the use of by-products and industrial co-products would strengthen the forestry sector and reduce pressure on EU forests [27]. Tropical forests also play a significant role in the waste utilization scenario. In Malaysia, researchers successfully partially replaced gasoline with bioethanol produced from oil palm and rubber-wood residues, which could contribute significantly to the economy and sustainability of the transportation sector [73]. In the Brazilian Amazon, the production of charcoal from wood residues is also an environmentally interesting alternative [99].

In addition to wood, forests provide other high-value products. In the Amazon, the rosewood tree (*Aniba rosaeodora* Ducke) éis a valuable source of essential oil rich in linalool, highly sought after by the perfume and cosmetics industries. However, predatory exploitation, characterized by clear-cutting since the 1920s [141], has led the species to be listed as endangered by Brazilian authorities, and it remains in this condition to this day [112]. The oil extraction was carried out exclusively from the wood, generating residues such as leaves and branches. Additionally, the exploitation area occurred in dense forests and resulted in the destruction of smaller individuals and the surrounding area, generating more biomass waste. Studies conducted in the 1970s [10] indicated the possibility of extracting essential oil from leaves and branches, thus avoiding the felling of trees. Other studies confirmed the feasibility of producing biomass from the regrowth of rosewood tree crowns, offering an alternative to obtain additional biomass without sacrificing the trees. Sustainable use of biomass is crucial to reduce waste, and in the case of rosewood, it could even eliminate waste, provided sustainable management practices are adopted to transform them into co-products, benefiting native biomes and their biodiversity.

The pulp and paper industries, connected to the forest-based sector, play a significant role in the global economy and are intrinsically linked to human life. However, due to their large scale, the production process generates considerable waste that threatens the environment [97]. In 2022, the production of various types of paper exceeded 518 million tons [50], with specificities in the production process but inherent waste generation, including primary and secondary sludge and black liquor. The effluents generated contain organic compounds from decomposed carbohydrates, lignin, extractives, and organochlorines resulting from bleaching stages that use chlorine-based products [132]. Black liquor, a valuable byproduct of the kraft pulping process, contains lignin and inorganic chemicals [66]. It is generated at a rate of 1.4–1.5 tons per ton of cellulose produced, representing a significant potential energy source of 250–500 MW [87]. The composition of black liquor varies depending on the raw material and processing methods, with lignin content ranging from 10 to 50% [8]. Despite its lower calorific value of 12–13 MJ/kg, black liquor is increasingly recognized as a renewable energy source and a platform for producing various materials [8, 87]. A potential application of lignin derived from black liquor is as a precursor for carbon-based materials. Some studies use it as a precursor for activated carbon employed in wastewater treatment [184] and electrochemically active materials essential in environmentally friendly supercapacitors [78]. Another application worth mentioning is the use of liquor as a macromolecule in biopolymer blends [20] or in the composition of remediation hydrogels [71] and phenol–formaldehyde resins [136].

2.1.3 Animal Wastes

According to estimates from the [50] *Food Outlook*, poultry meat production will reach 146 million tons, followed by pork at 123.3 million tons, beef at 77.2 million tons, and mutton at 17.3 million tons [50]. In the same period, milk production is expected to reach 978.5 million tons. The figures for fish production, including fishing and aquaculture, are estimated to reach 191.4 million tons (fresh weight). These data highlight the substantial contribution of animal production to biomass generation, with each sector responsible for creating waste that can be transformed into higher value-added products. Some examples are shown in Fig. 4.

Hydrolysates derived from poultry slaughterhouse by-products, for example, can be used in applications such as antioxidants, anti-diabetic, anti-inflammatory, anticoagulant, anti-anemic, as well as being cardioprotective, hepatoprotective, and neuroprotective [75]. Animal bones have been used as raw material for bone char, which offers an excellent cost–benefit ratio as a stabilizer of heavy metals (Cu, Zn, Pb,

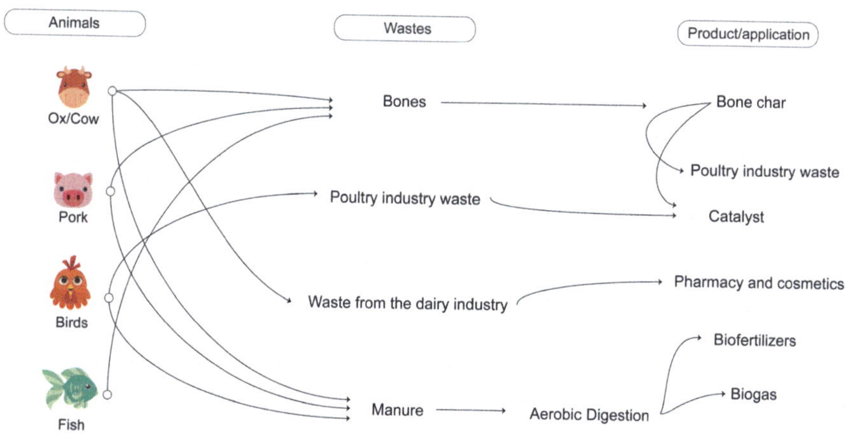

Fig. 4 Main animal sources, types of waste generated, and potential uses in various sustainable applications. *Source* Authors (2024)

and Cd) in soil, as well as promoting plant growth and reducing bioaccumulation in their roots [109]. In the case of fish bones, they are also used as a support for catalysts in biodiesel production processes [113]. The same biomass and eggshells were assessed in similar studies seeking alternatives to commercial catalysts for hydrocarbon production in bio-oil [150]. For nanomaterial purposes, fish bones have been considered a source for producing biphasic bioceramics based on calcium phosphate [32]. This type of biomaterial exhibits good biocompatibility, bioactivity, and osteoconductivity with implanted tissues [72].

For various animal farms, manure production is inherent, and it is commonly used for bioenergy generation and biofertilizers with the aid of biodigesters and anaerobic digestion processes [151]. The conversion of this waste through hydrothermal liquefaction can produce biocrude, with potential for energy input, and hydrochar, which can also be applied in energy production or serve as an adsorbent and soil conditioner [101]. There are also innovative studies using cattle manure to produce nano and microparticles as an alternative to chemical fertilizers [13]. This new product offers two main advantages over fertilizers: slower nutrient release, allowing for prolonged absorption by plants, and reduced nutrient loss in the soil. According to Hoyos-Sebá et al. [74], animal manure has clear potential for producing bioenergy, biomaterials, and organic acids. Still, there are obstacles and bottlenecks to be identified and overcome for its use, including biomass distribution and the presence of antibiotics in manure. In the dairy industry, waste can also be transformed into high-value products. Sar et al. [144] summarize those microbial processes, both aerobic and anaerobic, can utilize lactose, fat, and proteins inherent to these wastes. Additionally, this biomass has a high chemical oxygen demand. Thus, in anaerobic digestion, the waste can be converted into biogas. In aerobic cultivation, it can generate bioethanol, organic acids, enzymes, bacterial cellulose, and pigments.

3 Valorization of Agroindustrial Waste in the Pharmaceutical Industry

Using active ingredients extracted from agroindustrial waste is a strategic alternative to synthetic compounds, promoting independence from non-renewable sources in pharmaceutical production [125]. Replacing synthetic sources can reduce the environmental impacts resulting from the production of these compounds. Petrochemical processes used in the processing of synthetic drugs are energy-intensive and often generate new types of waste, which can sometimes be toxic [47, 181, 182]. On the other hand, the extraction of bioactive compounds from agroindustrial waste provides a more sustainable approach to recycling and a circular economy. Extracting bioactive compounds from agroindustrial waste allows the pharmaceutical sector to access low-cost raw materials and reduce operational expenses related to synthetic inputs [116]. Therefore, applying these waste materials presents an efficient solution to reduce environmental impacts and production costs,previously overlooked waste is transformed into bioactive sources of bioactives.

As consumers and regulators demand more environmental responsibility and sustainable products, the utilization of waste generates a positive impact on the corporate image of industries [118]. Companies that use waste as raw material can obtain environmental certifications, such as ISO 14001, which increases their appeal to consumers and investors who prioritize sustainability values in decision-making [106, 140]. Compliance with environmental regulations can also facilitate access to tax incentives and government subsidies for environmentally responsible practices [52]. The utilization of agroindustrial waste by the pharmaceutical industry becomes a source of innovation, as extracting active ingredients from these materials can lead to the development of new types of drugs distinct from traditional synthetic inputs.

3.1 Active Compounds

Active compounds are chemical substances that provide a medication's biochemical and therapeutic action [22]. They can be of natural origin, extracted from plant compounds such as plants, agroindustrial waste, and microorganisms, or of synthetic origin, produced in laboratories [92, 149]. In drug production, these active compounds are the substances responsible for performing preventive, palliative, or curative functions in the body [129]. The bioactive properties of agroindustrial waste are a significant source of active compounds for pharmaceutical production [6, 96]. Among the most common active compounds found in agroindustrial waste are flavonoids, polyphenols, alkaloids, fatty acids, and polysaccharides, which possess antioxidant, anti-inflammatory, and antimicrobial properties with broad applicability for human health [149].

Polyphenols and flavonoids are bioactive compounds commonly found in plant residues such as citrus and pineapple peels, apple and grape pomace, and mango seeds [6, 31, 64, 70]. The polyphenols and flavonoids present in these residues exhibit specifically antioxidant and anti-inflammatory properties [45, 48, 124]. These properties can be promising in the development of medications aimed at the prevention and treatment of chronic diseases, such as cancer and cardiovascular diseases, due to their ability to neutralize certain free radicals, inhibit cell proliferation, and reduce inflammation [9, 62, 167]. In addition to their antioxidant functions, flavonoids have also been linked to improvements in brain health, helping to modulate signaling pathways that influence synaptic plasticity [148, 167].

Another prominent bioactive compound group is alkaloids, found in residues from coffee and cocoa products [36, 138]. One of the most studied alkaloids is caffeine, which is widely known as an important central nervous system stimulant and is already being used in medications aimed at improving concentration, for example [36]. Theobromine, primarily found in residues of cocoa-based products, is also an important alkaloid with diuretic and vasodilatory properties, potential analgesic effects, and applications in research for treatments of circulatory problems [18, 28, 119]. Essential fatty acids also have important therapeutic functionalities and are found in residues from olive pomace, grape seeds, and other by-products from vegetable oil processing [67]. In addition, they play an important role in preventing cardiovascular diseases, as they contribute to reducing low-density lipoprotein cholesterol [40]. Due to their role in cell membrane balance and the regulation of inflammatory processes, these compounds have gained prominence in the pharmaceutical industry [67].

Found in various agroindustrial residues, such as cereal bran and fruit pomace, polysaccharides are compounds of great functional and nutritional importance [149, 168]. They are complex carbohydrates that play a prominent role in the pharmaceutical industry, exhibiting immunoregulatory, antioxidant, hypoglycemic, and antiviral activities [183]. Studies show that polysaccharides, such as starch and cellulose, have beneficial effects on human health, including prebiotic effects and immune system modulation [14]. Cellulose helps the body absorb nutrients and improves digestion [120]. Other polysaccharides, such as β-glucans found in oats, are associated with blood sugar control and cholesterol reduction [14]. In general, polysaccharides can be used as excipients in drug formulations, contributing to the stability of bioactive compounds [154]. Polysaccharides such as alginates and agar are used in gel and capsule formulations because of their thickening and gelling properties [84, 128]. The versatility of polysaccharides, like other bioactive compounds in agroindustrial residues, has increasingly attracted attention, especially in a context where waste conversion and sustainability are becoming essential.

3.2 *Bioactive Compound Extraction Processes: Extraction and Purification*

Bioactive compounds' extraction and purification techniques derived from agroindustrial residues are essential for ensuring efficient and safe chemical formulations. Various techniques are applied to maximize the high-quality extraction of these bioactive compounds, including solvent extraction, steam distillation, supercritical extraction, and other emerging technologies [5]. One of the most common and economical techniques is solvent extraction, which involves dissolving the compounds of interest using a suitable solvent [186]. The use of solvents such as hexane (nonpolar) and ethanol (polar) allows for the selective separation of bioactive compounds based on their polarities [155, 162]. Therefore, the choice of solvent is of great importance in ensuring extraction efficiency without compromising the compounds of interest. Although this technique is widely used, it has sustainability limitations. The use of traditional organic solvents, if not properly managed, may present environmental and health risks. Large volumes of solvents can generate substantial amounts of toxic waste, increase the costs associated with effluent treatment, and consume more energy in the process [155], which increases its carbon footprint and makes the technique less attractive from a sustainability perspective.

Supercritical extraction, which uses CO_2 (pressurized and heated until it reaches the supercritical state), is gaining attention for its ability to extract compounds without the need for toxic solvents. Additionally, the CO_2 used in the extraction can be recovered and recycled after the process [7, 90]. Meanwhile, steam distillation stands out for its efficiency in separating volatile compounds, such as extracting certain alkaloids and essential oils [12]. This technique is particularly useful for extracting bioactive compounds that have high decomposition and degradation temperatures,it uses steam to transfer the compounds of interest, which are then condensed and collected [166]. Factors such as extraction time, temperature, and the solid-to-liquid ratio significantly influence the purity and yield of the obtained extracts [16]. The subsequent purification of the obtained extracts involves chromatography and filtration, which remove impurities and concentrate the bioactive compounds [165]. It is crucial to go through the purification step to ensure that unwanted compounds are at levels lower than necessary to obtain the desired properties of the bioactive compounds required for potential pharmaceuticals [139].

4 Bioactive Potential of Agroindustrial Residues and Sustainability

The demand for effective medicines drives the pharmaceutical industry to explore alternative raw material sources, such as agro-residues, for the extraction of bioactive compounds [89]. Annually, the agroindustrial sector generates significant amounts of waste that negatively impacts public health, the environment, and socio-economic contexts [145]. Agroindustrial residues such as seeds, leaves, peels, and pomace from agricultural processing are prominent for bioactive compound extraction, with pharmacological properties widely documented in the literature [149]. The same authors explain that there is a vast potential for active ingredients derived from these wastes, most of which remain underexplored. Proteins, lipids, phenolic compounds, and polysaccharides are among the many bioactive molecules found in agroindustrial wastes [146, 149].

The extraction, identification, and evaluation of the potential of these bioactive compounds represent an important opportunity capable of promoting the utilization of various agroindustrial wastes [181, 182]. As social and environmental responsibility, along with the search for renewable raw materials, gain increasing relevance and become a necessity for a sustainable future, the use of agroindustrial wastes aligns with circular economy and sustainability goals [118, 142, 180]. In addition to offering medicinal benefits, bioactive compounds extracted from these materials contribute to reducing the accumulation of waste that previously had little added value [145]. Using these wastes in pharmaceutical production helps meet global environmental goals and strengthens the corporate image of industries adopting these practices, aligning them with innovation and sustainability trends.

Research related to the utilization of agroindustrial waste-based in the manufacture of new medications has intensified in recent years, driven by the abundance of bioactive compounds present in different plant parts [83, 102, 107]. These residual biomass, once discarded, have revealed great potential for the development of new therapies, primarily due to the presence of a variety of compounds such as carbohydrate molecules and pectins [33], as well as a wide range of compounds like flavonoids, phenolic acids [143], alkaloids, steroids, saponins, and quinine [159]. Each of these substances has characteristics that can be explored in various fields, including the pharmaceutical, cosmetic, and functional food industries, expanding the potential of utilizing these discarded materials.

Each of these compounds has been extensively studied for its therapeutic properties, including antioxidant activities and anticancer, anti-aging, and anti-inflammatory effects [17, 185]. For example, flavonoids and phenolic acids present in fruit and vegetable residues are widely studied for their antioxidant activities, making them promising in combating diseases related to oxidative stress and cellular aging [143]. Alkaloids and steroids, commonly found in medicinal plant residues, have been investigated for their anticancer and anti-inflammatory potential [159]. Additionally, compounds such as taxifolin have shown antitumor properties, further reinforcing the value of these wastes as renewable sources for discovering and developing new drugs [134]. Therefore, the therapeutic potential of agroindustrial residual biomass contributes to the research and formulation of new natural pharmacological agents, many of which exhibit bioactive activities of pharmaceutical interest. Reusing these by-products offers a sustainable alternative by transforming agroindustrial waste into sources of bioactive compounds, reinforcing circular economy principles, which aim to reuse waste in new production processes.

The growing demand for sustainable alternatives in the pharmaceutical industry has also driven the investigation of these bioactive compounds [11], such as polyphenols, gallic acid, and pyrogallol. For example, the bioconversion of by-products from the wood of fruit-bearing species has proven to be an effective technique for obtaining bioactive compounds that can be used to formulate new drugs. Processes such as pyrolysis, which converts lignocellulose into bio-oils, have been explored for their medicinal potential. Recent research demonstrated that the bio-oil derived from residual wood of *C. sinensis* showed promising antiviral activity against COVID-19 [172], highlighting the impact of these technologies in the search for innovative therapeutic solutions. In addition to wood, residues from other plant sources have also been extensively explored. For example, residual leaves of turmeric (*Curcuma longa* L.) have been investigated for their repellent and insecticidal effects, with no phytotoxic effects [179]. The reuse of these residues not only contributes to environmental sustainability but also expands the possibilities for medicinal application.

Diseases such as diabetes have also been the subject of studies involving agroindustrial residues. Grape pomace has proven effective in controlling blood sugar levels and is widely investigated for its bioactive properties [81]. Other residues, such as the stem bark of sibipiruna (*Poincianella pluviosa* (DC.) L.P. Queiroz) and the green peel of pistachio (*Pistacia vera* L.), have also been analyzed for their therapeutic potential in treating diabetes [82, 94]. In oncology, agroindustrial residues are being investigated for their therapeutic applications in alternative cancer treatments. For example, grape pomace (*Vitis vinifera* L.) and seeds have been studied to treat colorectal cancer [126]. Additionally, residues from the peel of cashew (*Anacardium occidentale* L.), black wattle (*Acacia mearnsii* De Wild.), and *barbatimão* (*Stryphnodendron adstringens* (Mart.) Coville) have shown antioxidant, antibacterial, and anticancer properties, making them promising for the treatment of gastric cancer [173].

These advances highlight how research on agroindustrial residues is expanding and contributing to the development of multiple products. Using agricultural byproducts brings therapeutic benefits and promotes sustainable practices by reducing waste and reusing residues rich in bioactive compounds. In this way, these materials, once discarded, are reintegrated into productive chains, aligning with the principles of the circular economy.

5 Connections, Challenges, and Barriers in Converting Agroindustrial Waste into Resources

The conversion of agroindustrial waste into high-value products is one of the fundamentals of the circular economy, aiming to reduce waste and maximize resource reuse. However, large-scale implementation of this transformation faces substantial challenges that hinder its full realization. One of the main difficulties lies in the lack of standardized methods capable of treating the vast diversity of waste generated in the agroindustrial sector, as each type of byproduct has specific chemical and physical characteristics. Therefore, these wastes require efficient treatment and utilization approaches [1].

Various treatment technologies can be applied to convert waste into energy or high-value products, and they can be classified into mechanical, thermochemical, and biochemical technologies [80]. However, each has its limitations. Mechanical processes, while helpful, may generate secondary waste streams that complicate further management. Chemical processes require large volumes of reagents, potentially compromising environmental sustainability. Lastly, biological processes face difficulties related to the need for specialized equipment, hindering their scalability and industrial efficiency [35]. This combination of technological challenges reveals the complexity of implementing solutions that transform waste into high-value products in a sustainable and economically viable manner.

Waste management in the agroindustry presents serious challenges in terms of infrastructure and regulation that directly impact the efficiency of the collection, disposal, and reuse of agricultural byproducts. Inadequate disposal practices, such as burning waste in the field, pose significant environmental concerns, resulting in the loss of essential nutrients and soil fertility degradation [80]. Moreover, this practice contributes to air pollution, causing irreversible damage to the environment.

The problem of waste collection is exacerbated by the lack of proper infrastructure, particularly in rural areas. The high cost and limited availability of machines for small and medium farmers are notable obstacles. Often, hiring waste management services is not feasible due to the high energy demands and the specificity of machinery for different types of crops, making it challenging to adopt these practices in smaller properties [1]. This limitation is further intensified when considering regional socioeconomic and cultural factors, which also influence the efficiency of waste management and the extraction of valuable bioproducts from these materials [131].

Solid support through public policies that encourage recycling and the reuse of waste is essential to overcome these barriers. Investing in innovative technologies that can efficiently and sustainably process and reuse these byproducts is a pressing need [43]. However, the mere existence of technologies is not enough. Clear and comprehensive regulatory frameworks are crucial to ensure that waste management is carried out in an environmentally safe and economically viable manner. Policies governing this area play a central role in the agricultural sector's feasibility of circular economy initiatives [46].

In this regard, innovative financing models, such as public–private partnerships and collaboration between the financial, industrial, and governmental sectors, can mobilize the resources needed to promote proper infrastructure and technology development [105]. In Brazil, initiatives such as the Programa Nacional de Resíduos Sólidos [National Solid Waste Program] promote circular economy practices and waste reduction through collaborations between governments, recycling cooperatives, and private companies. These collaborations are essential for driving significant investments, ensuring that waste management practices become more accessible, both financially and technologically, to small and medium farmers.

In this context, it is essential to recognize that the bioeconomy approach to agroindustrial waste valorization involves several key actors across different value chains. These actors may change depending on socioeconomic conditions, which can either facilitate or hinder waste valorization. The transition to a sustainable circular bioeconomy is complex and requires a systemic approach involving technological, organizational, and social innovations [44]. In the case of Malaysia, a positive example of government involvement in palm oil production can be observed. Through the implementation of regulatory licenses and certifications, the government establishes a set of principles and criteria for sustainable cultivation [68]. It would then be possible to extend these mechanisms to guide the use of residual biomass through biorefinery technologies, thereby creating an industry for bioproducts while minimizing waste impact. According to Liu et al. [100], it is also necessary to focus on promoting technology among agricultural producers. This way, we will have waste conversion technologies through research and technological development, which can be promoted through practical guides and financial incentives. Finally, this will spread the use of residual biomass and its newly generated products.

6 Final Considerations

Generally, we observe that the diversity of raw materials from numerous agro-industrial production chains holds significant potential for waste transformation and valorization. Advancing technological and scientific progress for utilizing these by-products is a crucial step toward a low-impact economy in harmony with finite resources. The examples presented in this chapter demonstrate that science is dedicated to pushing beyond established boundaries and exploring new practices and products derived from waste. However, for technological advancements to align with sustainable development, more concrete actions are needed from global decision-makers. Industries and governments must establish a harmonious relationship between economic development, natural resources, and human needs, moving beyond wealth concentration. We emphasize that utilizing waste for multiple products is an opportunity to strengthen an economy that contrasts with the infinite generation of goods and capital accumulation. In summary, this is a critical opportunity for the sustainability of key sectors of human activity, such as agriculture, livestock, forestry, forest extraction, and all industrial links in their supply chains.

References

1. Abdel-Shafy HI, Mansour MSM (2018) Solid waste issue: sources, composition, disposal, recycling, and valorization. Egypt J Pet 27(4):1275–1290. https://doi.org/10.1016/J.EJPE.2018.07.003
2. Acquah ISK, Quaicoe J, Gatsi JG (2024) Modelling circular economy capabilities and sustainable manufacturing practices for environmental performance: assessing linear (PLS-SEM) and non-linear (fsQCA) effects. Technol Forecast Soc Change 205:123501. https://doi.org/10.1016/J.TECHFORE.2024.123501
3. Adamu H, Bello U, Yuguda AU, Tafida UI, Jalam AM, Sabo A, Qamar M (2023) Production processes, techno-economic and policy challenges of bioenergy production from fruit and vegetable wastes. Renew Sustain Energy Rev 186:113686. https://doi.org/10.1016/j.rser.2023.113686
4. Adetunji AI, Oberholster PJ, Erasmus M (2023) From garbage to treasure: areview on biorefinery of organic solid wastes into valuable biobased products. Bioresour Technol Rep 24:101610. https://doi.org/10.1016/j.biteb.2023.101610
5. Agrawal RS, Nirmal NP (2024) Technologies for extraction of bioactive compounds and its applications. In: Sarkar T, Pati S (eds) Bioactive extraction and application in food and nutraceutical industries. Springer US, New York, NY, pp 1–20. https://doi.org/10.1007/978-1-0716-3601-5_1
6. Ahmed N (2023) Extraction of bioactive compounds from agro-industrial waste. In: Extraction of natural products from agro-industrial wastes: a green and sustainable approach. Elsevier, pp 131–142
7. Aili Q, Cui D, Li Y, Zhige W, Yongping W, Minfen Y, Dongbin L, Xiao R, Qiang W (2024) Composing functional food from agro-forest wastes: selectively extracting bioactive compounds using supercritical fluid extraction. Food Chem 455
8. Al-Kaabi Z, Pradhan R, Thevathasan N, Arku P, Gordon A, Dutta A (2018) Beneficiation of renewable industrial wastes from paper and pulp processing. AIMS Energy 6:880–907. https://doi.org/10.3934/energy.2018.5.880

9. AL-Khayri JM, Sahana GR, Nagella P, Joseph BV, Alessa FM, Al-Mssallem MQ (2022) Flavonoids as potential anti-inflammatory molecules: a review. Molecules 27
10. Araujo VCD, Corrêa GC, Maia JGS, Silva MLD, Gottlieb OR, Marx MC, Magalhães MT (1971) Óleos essenciais da Amazonia contendo linalol. Acta Amaz 1:45–47. https://doi.org/10.1590/1809-43921971013045
11. Ariwaodo CA, Olaniyan OF (2024) Fleshy fruit waste and the green chemistry of its conversion to valuable products for humans and animals. Food Chem Adv 4:100634. https://doi.org/10.1016/j.focha.2024.100634
12. Arun KB, Madhavan A, Sindhu R, Binod P, Pandey A, Sirohi R (2020) Remodeling agro-industrial and food wastes into value-added bioactives and biopolymers. Ind Crops Prod 154. https://doi.org/10.1016/j.indcrop.2020.112621
13. Aryanpour H, Naeini SAM, Ahmadian A (2017) Application of nano- and micro-sized particles of cattle manure on soybean growth. Environ Health Eng Manage 4:239–244. https://doi.org/10.15171/ehem.2017.33
14. Arzami AN, Ho TM, Mikkonen KS (2022) Valorization of cereal by-product hemicelluloses: fractionation and purity considerations. Food Res Int 151
15. Asyifah MR, Lu K, Ting HL, Zhang D (2014) Hidden potential of tropical fruit waste components as a useful source of remedy for obesity. J Agric Food Chem 62:3505–3516. https://doi.org/10.1021/jf5007352
16. Baglary GR, Kalita S, Islary A, Kumar S (2024) Sustainable extraction of bioactive compounds from aromatic plants and agro-food wastes for food preservation: a review. Biocatal Agric Biotechnol 61:103399. https://doi.org/10.1016/j.bcab.2024.103399
17. Bala S, Garg D, Sridhar K, Inbaraj BS, Singh R, Kamma S, Tripathi M, Sharma M (2023) Transformation of agro-waste into value-added bioproducts and bioactive compounds: micro/nano formulations and application in the agri-food-pharma sector. Bioeng 10(2):152. https://doi.org/10.3390/BIOENGINEERING10020152
18. Bartella L, Di Donna L, Napoli A, Siciliano C, Sindona G, Mazzotti F (2019) A rapid method for the assay of methylxanthines alkaloids: theobromine, theophylline and caffeine, in cocoa products and drugs by paper spray tandem mass spectrometry. Food Chem 278:261–266. https://doi.org/10.1016/J.FOODCHEM.2018.11.072
19. BarthHaas (2024) BarthHaas report 2023/2024 general. BarthHaas GmbH & Co, Nuremberg
20. Bass GF, Epps TH (2021) Recent developments towards performance-enhancing lignin-based polymers. Polym Chem 12:4130–4158. https://doi.org/10.1039/D1PY00694K
21. Ben-Othman S, Jõudu I, Bhat R (2020) Bioactives from agri-food wastes: present insights and future challenges. Molecules 25(3):510. https://doi.org/10.3390/molecules25030510
22. Braga MC, Vieira ECS, de Oliveira TF (2018) Curcuma longa L. leaves: characterization (bioactive and antinutritional compounds) for use in human food in Brazil. Food Chem 265:308–315. https://doi.org/10.1016/j.foodchem.2018.05.096
23. Brondízio ES (2004) From staple to fashion food: shifting cycles and shifting opportunities in the development of the Açaí Palm fruit economy in the Amazon estuary. In: Zarin DJ, Alavalapati JRR, Putz FE, Schmink M (eds) Working forests in the neotropics: conservation through sustainable management? Columbia University Press, pp 339–365. https://doi.org/10.7312/zari12906-021
24. Bufalino L, Guimarães AA, Silva BMDSE, De Souza RLF, De Melo ICNA, De Oliveira DNPS, Trugilho PF (2018) Local variability of yield and physical properties of açaí waste and improvement of its energetic attributes by separation of lignocellulosic fibers and seeds. J Renew Sustain Energy 10. https://doi.org/10.1063/1.5027232
25. Buller LS, Sganzerla WG, Lima MN, Muenchow KE, Timko MT, Forster-Carneiro T (2022) Ultrasonic pretreatment of brewers' spent grains for anaerobic digestion: biogas production for a sustainable industrial development. J Clean Prod 355:131802. https://doi.org/10.1016/j.jclepro.2022.131802
26. Cádiz-Gurrea MDLL, Villegas-Aguilar MDC, Leyva-Jiménez FJ, Pimentel-Moral S, Fernández-Ochoa Á, Alañón ME, Segura-Carretero A (2020) Revalorization of bioactive compounds from tropical fruit by-products and industrial applications by means of sustainable approaches. Food Res Int 138:109786. https://doi.org/10.1016/j.foodres.2020.109786

27. Camia A, Robert N, Jonsson R, Pilli R, García-Condado S, López-Lozano R, van der Velde M, Ronzon T, Gurría P, M'Barek R, Tamosiunas S, Fiore G, Araujo R, Hoepffner N, Marelli L, Giuntoli J (2018) Biomass production, supply, uses and flows in the European Union: first results from an integrated assessment. Publications Office of the European Union, Luxembourg
28. Camps-Bossacoma M, Pérez-Cano FJ, Franch À, Castell M (2018) Theobromine is responsible for the effects of cocoa on the antibody immune status of rats. J Nutr 148(3):464–471. https://doi.org/10.1093/JN/NXX056
29. Capanoglu E, Nemli E, Tomas-Barberan F (2022) Novel approaches in the valorization of agricultural wastes and their applications. J Agric Food Chem 70(23):6787–6804. https://doi.org/10.1021/ACS.JAFC.1C07104/ASSET/IMAGES/LARGE/JF1C07104_0001.JPEG
30. Cardoso D, Särkinen T, Alexander S, Amorim AM, Bittrich V, Celis M, Daly DC, Fiaschi P, Funk VA, Giacomin LL, Goldenberg R, Heiden G, Iganci J, Kelloff CL, Knapp S, Cavalcante De Lima H, Machado AFP, Dos Santos RM, Mello-Silva R, Michelangeli FA, Mitchell J, Moonlight P, De Moraes PLR, Mori SA, Nunes TS, Pennington TD, Pirani JR, Prance GT, De Queiroz LP, Rapini A, Riina R, Rincon CAV, Roque N, Shimizu G, Sobral M, Stehmann JR, Stevens WD, Taylor CM, Trovó M, Van Den Berg C, Van Der Werff H, Viana PL, Zartman CE, Forzza RC (2017) Amazon plant diversity revealed by a taxonomically verified species list. Proc Natl Acad Sci USA 114:10695–10700. https://doi.org/10.1073/pnas.1706756114
31. Chakka AK, Babu AS (2022) Bioactive compounds of winery by-products: extraction techniques and their potential health benefits. Appl Food Res 2(1):100058. https://doi.org/10.1016/j.afres.2022.100058
32. Corrêa THA, Holanda JNF (2019) Fish bone as a source of raw material for synthesis of calcium phosphate. Mat Res 22:e20190486. https://doi.org/10.1590/1980-5373-mr-2019-0486
33. Cruz Reina LJ, Durán-Aranguren DD, Forero-Rojas LF, Tarapuez-Viveros LF, Durán-Sequeda D, Carazzone C, Sierra R (2022) Chemical composition and bioactive compounds of cashew (Anacardium occidentale) apple juice and bagasse from Colombian varieties. Heliyon 8(5):e09528. https://doi.org/10.1016/J.HELIYON.2022.E09528
34. Da Silva MACN, Tessmann JW, Borges KRA, Wolff LAS, Botelho FD, Vieira LA, Morgado-Diaz JA, Franca TCC, Barbosa MDCL, Nascimento MDDSB, Rocha MR, De Carvalho JE (2023) Açaí (Euterpe oleracea Mart.) seed oil exerts a cytotoxic role over colorectal cancer cells: insights of Annexin A2 regulation and molecular modeling. Metabolites 13:789. https://doi.org/10.3390/metabo13070789
35. Dalbanjan NP, Eelager MP, Korgaonkar K, Gonal BN, Kadapure AJ, Arakera SB, Praveen Kumar SK (2024) Descriptive review on conversion of waste residues into valuable bionanocomposites for a circular bioeconomy. Nano Struct Nano Objects 39:101265. https://doi.org/10.1016/J.NANOSO.2024.101265
36. De Melo Pereira GV, de Carvalho Neto DP, Magalhães Júnior AI, do Prado FG, Pagnoncelli MGB, Karp SG, Soccol CR (2020) Chemical composition and health properties of coffee and coffee by-products. In: Advances in food and nutrition research. Academic Press Inc., pp 65–96
37. De Oliveira JAR, Da Conceição AC, Da Silva Martins LH, Moreira DKT, Passos MF, Komesu A (2021) Evaluation of the technological potential of four wastes from Amazon fruit industry in glucose and ethanol production. J Food Process Eng 44:e13610. https://doi.org/10.1111/jfpe.13610
38. De Oliveira PRS, Trugilho PF, de Oliveira TJP (2022) Briquettes of acai seeds: characterization of the biomass and influence of the parameters of production temperature and pressure in the physical-mechanical and energy quality. Environ Sci Pollut Res 29. https://doi.org/10.1007/s11356-021-15847-6
39. de Souza LKC, Gonçalves AAS, Queiroz LS, Chaar JS, da Rocha Filho GN, da Costa CEF (2020) Utilization of acai stone biomass for the sustainable production of nanoporous carbon for CO2 capture. Sustain Mater Technol 25. https://doi.org/10.1016/j.susmat.2020.e00168

40. Demirbas A (2008) Oils from hazelnut shell and hazelnut kernel husk for biodiesel production. Energy Sources Part A Recovery Utilization Environ Effects 30(20). https://doi.org/10.1080/15567030701457467
41. Diaz-Bonilla E, Echeverria R (2020) Duality, urbanization, and modernization of agrifood systems in Latin America and the Caribbean. International Food Policy Research Institute, Washington, DC. https://doi.org/10.2499/9780896293830_06
42. Ding Z, Ge Y, Sar T, Kumar V, Harirchi S, Binod P, Sirohi R, Sindhu R, Wu P, Lin F, Zhang Z, Taherzadeh MJ, Awasthi MK (2023) Valorization of tropical fruits waste for production of commercial biorefinery products—a review. Biores Technol 374:128793. https://doi.org/10.1016/j.biortech.2023.128793
43. Divyabharathi R, Kalidasan B, Sakthi Suriya SSR, Chinnasamy S (2024) Recent advances in sustainable agro residue utilisation, barriers and remediation for environmental management: present insights and future challenges. Ind Crops Prod 216:118790. https://doi.org/10.1016/J.INDCROP.2024.118790
44. Donner M, Erraach Y, López-i-Gelats F, Manuel-i-Martin J, Yatribi T, Radić I et al (2022) Circular bioeconomy for olive oil waste and by-product valorisation: actors' strategies and conditions in the Mediterranean area. J Environ Manage 321:115836. https://doi.org/10.1016/j.jenvman.2022.115836
45. Dulf FV, Vodnar DC, Dulf EH, Pintea A (2017) Phenolic compounds, flavonoids, lipids and antioxidant potential of apricot (Prunus armeniaca L.) pomace fermented by two filamentous fungal strains in solid state system. Chem Cent J 11(1). https://doi.org/10.1186/s13065-017-0323-z
46. Duque-Acevedo M, Belmonte-Ureña LJ, Cortés-García FJ, Camacho-Ferre F (2020) Agricultural waste: review of the evolution, approaches and perspectives on alternative uses. Glob Ecol Conserv 22:e00902. https://doi.org/10.1016/J.GECCO.2020.E00902
47. Emenike EC, Emmanuel SS, Iwuozor KO, Okwu KC, Adeniyi AG (2024) Plant biomass materials in petrochemical application. In: Plant biomass applications. Elsevier, pp 351–383
48. Estrada-Gil L, Contreras-Esquivel JC, Flores-Gallegos C, Zugasti-Cruz A, Govea-Salas M, Mata-Gómez MA, Rodríguez-Herrera R, Ascacio-Valdés JA (2022) Recovery of bioactive ellagitannins by ultrasound/microwave-assisted extraction from Mexican Rambutan Peel (*Nephelium lappaceum* L.). Molecules 27(5). https://doi.org/10.3390/molecules27051592
49. Eurostat (2023) Forests, forestry and logging. Statistics Explained. https://ec.europa.eu/eurostat/statistics-explained/index.php?title=Forests,_forestry_and_logging. Accessed 12 Dec 2024
50. FAO (2024) Food outlook—biannual report on global food markets. FAO, Rome. https://doi.org/10.4060/cd1158en
51. Farcaş AC, Socaci SA, Nemeş SA, Pop OL, Coldea TE, Fogarasi M, Biriş-Dorhoi ES (2022) An update regarding the bioactive compound of cereal by-products: health benefits and potential applications. Nutrients 14:3470. https://doi.org/10.3390/nu14173470
52. Florek-Paszkowska AK, Hoyos-Vallejo CA (2023) A comprehensive bibliometric analysis and future research directions in the nexus of sustainable business practices and turnover intention. Cleaner Responsible Consumption 11
53. Fontana KB, Chaves ES, Sanchez JDS, Watanabe ERLR, Pietrobelli JMTA, Lenzi GG (2016) Textile dye removal from aqueous solutions by malt bagasse: isotherm, kinetic and thermodynamic studies. Ecotoxicol Environ Saf 124:329–336. https://doi.org/10.1016/j.ecoenv.2015.11.012
54. Food and Agriculture Organization of the United Nations—FAO (2013) Agricultural and forest residues: generation, utilization and availability. www.fao.org/docrep/006/AD576E/ad576e00.pdf. Accessed 18 Sept 2024
55. Food and Agriculture Organization of the United Nations—FAO (2010) Global forest resources assessment. FAO Forestry Paper, Rome
56. Food and Agriculture Organization of the United Nations—FAO (2022) The State of the World's Forests 2022. Forest pathways for green recovery and building inclusive, resilient and sustainable economies. Rome, FAO. https://doi.org/10.4060/cb9360en

57. Food and Agriculture Organization of the United Nations—FAO (2023) FAOSTAT: crops and livestock products. https://www.fao.org/faostat/en/#data/FO. Accessed 18 Oct 2024
58. Food and Agriculture Organization of the United Nations—FAO (2023) FAOSTAT: forestry production and trade. https://www.fao.org/faostat/en/#data/FO. Accessed 18 Sept 2024
59. Food and Agriculture Organization of the United Nations—FAO (2024) Land statistics 2001–2022 global, regional and country trends. FAO, Rome
60. Food and Agriculture Organization of the United Nations—FAO (2024) Food outlook—biannual report on global food markets. FAO, Rome. https://doi.org/10.4060/cd1158en
61. Food and Agriculture Organization of the United Nations—FAO (2024) Major tropical fruits market review 2023. Rome
62. González-Gallego J, García-Mediavilla MV, Sánchez-Campos S, Tuñó MJ (2010) Fruit polyphenols, immunity and inflammation. Br J Nutr 104
63. Gutiérrez-Grijalva EP, Cabanillas-Bojórquez LA, Vazquez-Olivo G, Heredia JB (2021) Cereal by-products as source of dietary fiber. In: Food loss and waste reduction. Apple Academic Press, Boca Raton, pp 161–174. https://doi.org/10.1201/9781003083900-9
64. Haida Z, Ab Ghani S, Juju Nakasha J, Hakiman M (2022) Determination of experimental domain factors of polyphenols, phenolic acids and flavonoids of lemon (Citrus limon) peel using two-level factorial design: determination of experimental domain factors. Saudi J Biol Sci 29(1):574–582. https://doi.org/10.1016/j.sjbs.2021.09.022
65. Han JW, Ruiz-Garcia L, Qian JP, Yang XT (2018) Food packaging: a comprehensive review and future trends. Compr Rev Food Sci Food Saf 17(4):860–877. https://doi.org/10.1111/1541-4337.12343
66. Harisankar S, Vinu R (2023) Platform chemicals from hardwood black liquor *via* hydrothermal liquefaction: influence of process conditions on product yields and quality. Sustain Energy Fuels 7:4423–4441. https://doi.org/10.1039/D3SE00308F
67. Hasnat A, Moheman A, Usmani MA, Bhawani SA (2023) Extraction of fatty acids from agro-industrial waste and its significance. In: Extraction of natural products from agro-industrial wastes: a green and sustainable approach. Elsevier, pp 233–242
68. Hassan MA, Farid MAA, Zakaria MR, Ariffin H, Andou Y, Shirai Y (2024) Palm oil expansion in Malaysia and its countermeasures through policy window and biorefinery approach. Environ Sci Policy 153:103671. https://doi.org/10.1016/j.envsci.2024.103671
69. Hatami T, Cidreira ACM, Linan LZ, Rodrigues J, Pinheiro IF, Viganó J, Chinelatto MA, Jardini AL, Martínez J, Mei LHI (2024) Valorization of acai bagasse into cellulose nanofibers/sodium alginate aerogels. Waste Biomass Valor 15:2451–2467. https://doi.org/10.1007/s12649-023-02281-7
70. He Y, Sun Z, Bai JY, Zhang Y, Qian Y, Zhao X, Chen S (2023) Citrus peel polyphenols alleviate intestinal inflammation in mice with dextran sulfate sodium-induced acute colitis. Heliyon 9(7). https://doi.org/10.1016/j.heliyon.2023.e18137
71. Heo JW, Kim MS, Oh DH, Kim YS (2024) Innovative construction of amine-functionalized lignin-based hydrogel for ultrafast and selective dye remediation. Chem Eng J 498:155506. https://doi.org/10.1016/j.cej.2024.155506
72. Hong Y, Fan H, Li B, Guo B, Liu M, Zhang X (2010) Fabrication, biological effects, and medical applications of calcium phosphate nanoceramics. Mater Sci Eng R Rep 70:225–242. https://doi.org/10.1016/j.mser.2010.06.010
73. Hossain N, Mahlia TMI, Miskat MI, Chowdhury T, Barua P, Chowdhury H, Nizamuddin S, Ahmad NB, Zaharin NAB, Mazari SA, Soudagar MEM (2021) Bioethanol production from forest residues and life cycle cost analysis of bioethanol-gasoline blend on transportation sector. J Environ Chem Eng 9:105542. https://doi.org/10.1016/j.jece.2021.105542
74. Hoyos-Sebá JJ, Arias NP, Salcedo-Mendoza J, Aristizábal-Marulanda V (2024) Animal manure in the context of renewable energy and value-added products: a review. Chem Eng Process Process Intensification 196:109660. https://doi.org/10.1016/j.cep.2023.109660
75. Ibarz-Blanch N, Alcaide-Hidalgo JM, Cortés-Espinar AJ, Albi-Puig J, Suárez M, Mulero M, Morales D, Bravo FI (2023) Chicken slaughterhouse by-products: a source of protein hydrolysates to manage non-communicable diseases. Trends Food Sci Technol 139:104125. https://doi.org/10.1016/j.tifs.2023.104125

76. Instituto Brasileiro de Geografia e Estatística—IBGE (2024) PAM: Produção Agrícola Municipal. Rio de Janeiro
77. International Energy Agency—IEA (2012) Global wood pellet industry market and trade study. IEA Task 40.av D, Kumar P, Panesar PS, Bunkar DS, Mishra D, Chopra HK (2016) Comparative study on conventional, ultrasonication and microwave assisted extraction of γ-oryzanol from rice bran. J Food Sci Technol 53(4). https://doi.org/10.1007/s13197-016-2175-2
78. Izaguirre N, Alberro M, Erdocia X, Labidi J (2024) Kraft and organosolv lignin-activated carbon composites for supercapacitor electrode materials. J Energy Storage 80:110386. https://doi.org/10.1016/j.est.2023.110386
79. Jellason NP, Robinson EJZ, Chapman ASA, Neina D, Devenish AJM, Po JYT, Adolph B (2021) A systematic review of drivers and constraints on agricultural expansion in Sub-Saharan Africa. Land 10:332. https://doi.org/10.3390/land10030332
80. Kapoor R, Ghosh P, Kumar M, Sengupta S, Gupta A, Kumar SS, Vijay V, Kumar V, Kumar Vijay V, Pant D (2020) Valorization of agricultural waste for biogas based circular economy in India: a research outlook. Bioresour Technol 304:123036. https://doi.org/10.1016/J.BIORTECH.2020.123036
81. Kato-Schwartz CG, Corrêa RCG, de Souza LD, de Sá-Nakanishi AB, de Almeida GG, Seixas FAV, Haminiuk CWI, Barros L, Ferreira ICFR, Bracht A, Peralta RM (2020) Potential anti-diabetic properties of Merlot grape pomace extract: an in vitro, in silico and in vivo study of α-amylase and α-glucosidase inhibition. Food Res Int 137:109462. https://doi.org/10.1016/J.FOODRES.2020.109462
82. Kato-Schwartz CG, Babeto De Sá-Nakanishi A, Guidi AC, De Almeida Gonçalves G, Bueno FG, Petroncine B, Zani M, Carlos J, De Mello P, Sérgio P, Bueno A, Augusto F, Seixas V, Bracht A, Peralta RM (2020) Carbohydrate digestive enzymes are inhibited by Poincianella pluviosa stem bark extract: relevance on type 2 diabetes treatment. Clin Phytoscience 6(1):1–11 (61). https://doi.org/10.1186/S40816-020-00177-W
83. Kaul K, Rajauria G, Singh R (2024) Valorization of agro-industrial waste for pectinase production and its influence on circular economy. Food Bioprod Process 148:141–153. https://doi.org/10.1016/J.FBP.2024.09.008
84. Kavoosi G, Derakhshan M, Salehi M, Rahmati L (2018) Microencapsulation of zataria essential oil in agar, alginate and carrageenan. Innov Food Sci Emerg Technol 45:418–425. https://doi.org/10.1016/j.ifset.2017.12.010
85. Kelley LC, Peluso NL, Carlson KM, Afiff S (2020) Circular labor migration and land-livelihood dynamics in Southeast Asia's concession landscapes. J Rural Stud 73:21–33. https://doi.org/10.1016/j.jrurstud.2019.11.019
86. Kiehbadroudinezhad M, Hosseinzadeh-Bandbafha H, Karimi K, Madadi M, Chisti Y, Peng W, Liu D, Tabatabaei M, Aghbashlo M (2023) Production of chemicals and utilities in-house improves the environmental sustainability of phytoplankton-based biorefinery. Sci Total Environ 899:165751. https://doi.org/10.1016/j.scitotenv.2023.165751
87. Kim C-H, Lee J-Y, Park S-H, Moon S-O (2019) Global trends and prospects of black liquor as bioenergy. J Korea TAPPI 51:3–15. https://doi.org/10.7584/JKTAPPI.2019.10.51.5.3
88. Kordi M, Farrokhi N, Pech-Canul MI, Ahmadikhah A (2024) Rice husk at a glance: from agro-industrial to modern applications. Rice Sci 31:14–32. https://doi.org/10.1016/j.rsci.2023.08.005
89. Kotwani A, Joshi J, Kaloni D (2021) Pharmaceutical effluent: a critical link in the interconnected ecosystem promoting antimicrobial resistance. Environ Sci Pollut Res 28(25):32111–32124. https://doi.org/10.1007/s11356-021-14178-w
90. Kumar P, Yadav D, Kumar P, Panesar PS, Bunkar DS, Mishra D, Chopra HK (2016) Comparative study on conventional, ultrasonication and microwave assisted extraction of γ-oryzanol from rice bran. J Food Sci Technol 53(4). https://doi.org/10.1007/s13197-016-2175-2
91. Kumar P (2022) Indian Council of Agricultural Research—ICAR. ICAR Annual Report 2022–23, New Delhi

92. Kumar S, Chauhan N, Tyagi B, Yadav P, Samanta AK, Tyagi AK (2023) Exploring bioactive compounds and antioxidant properties of twenty-six Indian medicinal plant extracts: a correlative analysis for potential therapeutic insights. Food Humanity 1:1670–1679. https://doi.org/10.1016/j.foohum.2023.11.016
93. Kumar V, Vangnai AS, Sharma N, Kaur K, Chakraborty P, Umesh M, Singhal B, Utreja D, Carrasco EU, Andler R, Awasthi MK, Taherzadeh MJ (2023) Bioengineering of biowaste to recover bioproducts and bioenergy: a circular economy approach towards sustainable zero-waste environment. Chemosphere 319:138005. https://doi.org/10.1016/J.CHEMOSPHERE.2023.138005
94. Lalegani S, Ahmadi Gavlighi H, Azizi MH, Amini Sarteshnizi R (2018) Inhibitory activity of phenolic-rich pistachio green hull extract-enriched pasta on key type 2 diabetes relevant enzymes and glycemic index. Food Res Int 105:94–101. https://doi.org/10.1016/J.FOODRES.2017.11.003
95. Lamb WF, Wiedmann T, Pongratz J, Andrew R, Crippa M, Olivier JGJ, Wiedenhofer D, Mattioli G, Khourdajie AA, House J, Pachauri S, Figueroa M, Saheb Y, Slade R, Hubacek K, Sun L, Ribeiro SK, Khennas S, De La Rue Du Can S, Chapungu L, Davis SJ, Bashmakov I, Dai H, Dhakal S, Tan X, Geng Y, Gu B, Minx J (2021) A review of trends and drivers of greenhouse gas emissions by sector from 1990 to 2018. Environ Res Lett 16:073005. https://doi.org/10.1088/1748-9326/abee4e
96. Leite P, Belo I, Salgado JM (2021) Co-management of agro-industrial wastes by solid-state fermentation for the production of bioactive compounds. Ind Crops Prod 172. https://doi.org/10.1016/j.indcrop.2021.113990
97. Liang X, Xu Y, Yin L, Wang R, Li P, Wang J, Liu K (2023) Sustainable utilization of pulp and paper wastewater. Water 15:4135. https://doi.org/10.3390/w15234135
98. Lima RS, De Carvalho APA, Conte-Junior CA (2023) Health from Brazilian Amazon food wastes: bioactive compounds, antioxidants, antimicrobials, and potentials against cancer and oral diseases. Crit Rev Food Sci Nutr 63:12453–12475. https://doi.org/10.1080/10408398.2022.2101983
99. Lima MDR, Simetti R, de Assis MR, Trugilho PF, de Carneiro ACO, Bufalino L, Hein PRG, de Protásio TP (2020) Charcoal of logging wastes from sustainable forest management for industrial and domestic uses in the Brazilian Amazonia. Biomass Bioenergy 142:105804. https://doi.org/10.1016/j.biombioe.2020.105804
100. Liu H, Long J, Zhang K, Li M, Zhao D, Song D et al (2024) Agricultural biomass/waste-based materials could be a potential adsorption-type remediation contributor to environmental pollution induced by pesticides—a critical review. Sci Total Environ 946:174180. https://doi.org/10.1016/j.scitotenv.2024.174180
101. Liu Q, Kong G, Zhang G, Cao T, Wang K, Zhang X, Han L (2023) Recent advances in hydrothermal liquefaction of manure wastes into value-added products. Energy Convers Manage 292:117392. https://doi.org/10.1016/j.enconman.2023.117392
102. Lopes FC, Ligabue-Braun R (2021) Agro-industrial residues: eco-friendly and inexpensive substrates for microbial pigments production. Front Sustain Food Syst 5:589414. https://doi.org/10.3389/FSUFS.2021.589414/BIBTEX
103. Machado CG, Winroth MP, Ribeiro da Silva EHD (2020) Sustainable manufacturing in Industry 4.0: an emerging research agenda. Int J Prod Res 58(5):1462–1484. https://doi.org/10.1080/00207543.2019.1652777
104. Mahato N, Sharma K, Sinha M, Cho MH (2018) Citrus waste derived nutra-/pharmaceuticals for health benefits: current trends and future perspectives. J Funct Foods 40:307–316. https://doi.org/10.1016/j.jff.2017.11.015
105. Makepa DC, Chihobo CH (2024) Barriers to commercial deployment of biorefineries: a multi-faceted review of obstacles across the innovation chain. Heliyon 10(12):e32649. https://doi.org/10.1016/J.HELIYON.2024.E32649
106. Malay OE, Aubinet S (2021) Improving government and business coordination through the use of consistent SDGs indicators. A comparative analysis of national (Belgian) and business (pharma and retail) sustainability indicators. Ecol Econ 184. https://doi.org/10.1016/j.ecolecon.2021.106991

107. Mármol I, Quero J, Ibarz R, Ferreira-Santos P, Teixeira JA, Rocha CMR, Pérez-Fernández M, García-Juiz S, Osada J, Martín-Belloso O, Rodríguez-Yoldi MJ (2021) Valorization of agro-food by-products and their potential therapeutic applications. Food Bioprod Process 128:247–258. https://doi.org/10.1016/J.FBP.2021.06.003
108. Meena PK, Awale SD, Kumar S, Kumar D, Kumar N (2024) Thermochemical conversion of agricultural residue for the production of hydrogen, methane, and biofuels: a comprehensive overview. Ind Crops Prod 221:119340. https://doi.org/10.1016/j.indcrop.2024.119340
109. Mei H, Huang W, Wang Y, Xu T, Zhao L, Zhang D, Luo Y, Pan X (2022) One stone two birds: bone char as a cost-effective material for stabilizing multiple heavy metals in soil and promoting crop growth. Sci Total Environ 840:156163. https://doi.org/10.1016/j.scitotenv.2022.156163
110. Miassi YE, Dossa KF (2024) Circular economy initiatives for forest-based bioeconomy: harnessing the potential of non-wood biomaterials. Waste Manage Bull 2:270–278. https://doi.org/10.1016/j.wmb.2024.05.006
111. Millati R, Cahyono RB, Ariyanto T, Azzahrani IN, Putri RU, Taherzadeh MJ (2019) Agricultural, industrial, municipal, and forest wastes. in: Sustainable resource recovery and zero waste approaches. Elsevier, pp 1–22. https://doi.org/10.1016/B978-0-444-64200-4.00001-3
112. Ministério do Meio Ambiente [Ministry of the Environment]—MMA (2022) Ordinance No. 148, of June 7, 2022. Official Gazette of the Union, Brasília, DF
113. Mohebolkhames E, Kazemeini M, Sadjadi S, Tamtaji M (2024) Synthesis and characterization of K_2CO_3 over Salmon-fishbone catalyst, kinetics and DFT analyses and biodiesel catalytic production from castor oil. J Mol Struct 1302:137426. https://doi.org/10.1016/j.molstruc.2023.137426
114. Muñoz P, Letelier V, Zamora D, Morales MP (2020) Feasibility of using paper pulp residues into fired clay bricks. J Clean Prod 262:121464. https://doi.org/10.1016/j.jclepro.2020.121464
115. Nasrin TAA, Matin MdA (2017) Valorization of vegetable wastes. In: Anal AK (ed) Food processing by-products and their utilization, 1st edn. Wiley, pp 53–88
116. Nath PC, Sharma R, Debnath S, Nayak PK, Roy R, Sharma M, Inbaraj BS, Sridhar K (2024) Recent advances in production of sustainable and biodegradable polymers from agro-food waste: applications in tissue engineering and regenerative medicines. Int J Biol Macromol 259. https://doi.org/10.1016/j.ijbiomac.2023.129129
117. Nautiyal H, Goel V (2021) Sustainability assessment: metrics and methods. Methods Sustain Sci Assess Prioritization Improv Des Optim 27–46. https://doi.org/10.1016/B978-0-12-823987-2.00017-9
118. Nesterov D, Barrera-Martínez I, Martínez-Sánchez C, Sandoval-González A, Bustos E (2024) Approaching the circular economy: biological, physicochemical, and electrochemical methods to valorize agro-industrial residues, wastewater, and industrial wastes. J Environ Chem Eng 12(5). https://doi.org/10.1016/j.jece.2024.113335
119. Neufingerl N, Zebregs YEMP, Schuring EAH, Trautwein EA (2013) Effect of cocoa and theobromine consumption on serum HDL-cholesterol concentrations: a randomized controlled trial. Am J Clin Nutr 97(6):1201–1209. https://doi.org/10.3945/AJCN.112.047373
120. Nsor-Atindana J, Zhou YX, Saqib MN, Chen M, Douglas Goff H, Ma J, Zhong F (2020) Enhancing the prebiotic effect of cellulose biopolymer in the gut by physical structuring via particle size manipulation. Food Res Int 131. https://doi.org/10.1016/j.foodres.2019.108935
121. Ogbu C, Okey SN (2023) Agro-industrial waste management: the circular and bioeconomic perspective. In: Ahmad F, Sultan M (eds) Agricultural waste—new insights. IntechOpen. https://doi.org/10.5772/intechopen.109181
122. Oliveira PRS, Setter C, Sousa NG, Cardoso CR, Trugilho PF, Oliveira TJP (2023) Kinetic and thermodynamic analysis of açaí seeds and insights into bio-oil optimization and composition. J Therm Anal Calorim 148:13427–13439. https://doi.org/10.1007/s10973-023-12645-9
123. Oliveira DNPS, Matos LC, Souza TM, Aráujo ES, Silva MG, Cabral CPT, Mendes LM, Rocha QS, Bufalino L (2024) Methods for separating the lignocellulosic fibers from the açaí pulping waste: quality for kraft pulping. Cerne 30:103376. https://doi.org/10.1590/01047760202430013376

124. Paes J, Dotta R, Barbero GF, Martínez J (2014) Extraction of phenolic compounds and anthocyanins from blueberry (Vaccinium myrtillus L.) residues using supercritical CO2 and pressurized liquids. J Supercrit Fluids 95. https://doi.org/10.1016/j.supflu.2014.07.025
125. Panzella L, Moccia F, Nasti R, Marzorati S, Verotta L, Napolitano A (2020) Bioactive phenolic compounds from agri-food wastes: an update on green and sustainable extraction methodologies. Front Nutr 7
126. Pérez-Ortiz JM, Alguacil LF, Salas E, Hermosín-Gutiérrez I, Gómez-Alonso S, González-Martín C (2019) Antiproliferative and cytotoxic effects of grape pomace and grape seed extracts on colorectal cancer cell lines. Food Sci Nutr 7(9):2948–2957. https://doi.org/10.1002/FSN3.1150
127. Pessoa JDC, Silva PVDS (2007) Effect of temperature and storage on açai (Euterpe oleracea) fruit water uptake: simulation of fruit transportation and pre-processing. Fruits 62:295–302. https://doi.org/10.1051/fruits:2007025
128. Putra NR, Fajriah S, Qomariyah L, Dewi AS, Rizkiyah DN, Irianto I, Rusmin D, Melati M, Trisnawati NW, Darwati I, Arya NN (2024) Exploring the potential of Ulva Lactuca: emerging extraction methods, bioactive compounds, and health applications – A perspective review. S Afr J Chem Eng 47:233–245. https://doi.org/10.1016/j.sajce.2023.11.017
129. Quagliariello V, Iaffaioli RV, Falcone M, Ferrari G, Pataro G, Donsì F (2016) Effect of pulsed electric fields—assisted extraction on anti-inflammatory and cytotoxic activity of brown rice bioactive compounds. Food Res Int 87:115–124. https://doi.org/10.1016/j.foodres.2016.07.005
130. Queiroz LS, De Souza LKC, Thomaz KTC, Leite Lima ET, Da Rocha Filho GN, Do Nascimento LAS, De Oliveira Pires LH, Faial KDCF, Da Costa CEF (2020) Activated carbon obtained from Amazonian biomass tailings (acai seed): modification, characterization, and use for removal of metal ions from water. J Environ Manage 270:110868. https://doi.org/10.1016/j.jenvman.2020.110868
131. Rame R, Purwanto P, Sudarno S (2023) Biotechnological approaches in utilizing agro-waste for biofuel production: an extensive review on techniques and challenges. Bioresour Technol Rep 24:101662. https://doi.org/10.1016/J.BITEB.2023.101662
132. Ramos MDN, Rangel AS, Azevedo KS, Melo MGB, Oliveira MC, Watanabe CMU, Pereira FF, Silva CM, Aguiar A (2022) Characteristics and treatment of Brazilian pulp and paper mill effluents: a review. Environ Monit Assess 194:651. https://doi.org/10.1007/s10661-022-10331-1
133. Ravindran B, Karmegam N, Yuvaraj A, Thangaraj R, Chang SW, Zhang Z, Kumar Awasthi M (2021) Cleaner production of agriculturally valuable benignant materials from industry generated bio-wastes: a review. Bioresour Technol 320:124281. https://doi.org/10.1016/J.BIORTECH.2020.124281
134. Razak S, Afsar T, Ullah A, Almajwal A, Alkholief M, Alshamsan A, Jahan S (2018) Taxifolin, a natural flavonoid interacts with cell cycle regulators causes cell cycle arrest and causes tumor regression by activating Wnt/ β-catenin signaling pathway. BMC Cancer 18(1):1–18. https://doi.org/10.1186/S12885-018-4959-4/FIGURES/11
135. Roberts L, Georgiou N, Hassan AM (2023) Investigating biodiversity and circular economy disclosure practices: insights from global firms. Corp Soc Responsib Environ Manage 30(3):1053–1069. https://doi.org/10.1002/CSR.2402
136. Rodrigues JS, De Freitas ADSM, Maciel CC, Mendes SF, Diment D, Balakshin M, Botaro VR (2023) Selection of kraft lignin fractions as a partial substitute for phenol in synthesis of phenolic resins: structure-property correlation. Ind Crops Prod 191:115948. https://doi.org/10.1016/j.indcrop.2022.115948
137. Rosen MA, Kishawy HA (2012) Sustainable manufacturing and design: concepts, practices and needs. Sustain 4(2):154–174. https://doi.org/10.3390/SU4020154
138. Saadi Ali H, Yardım Y (2024) Simultaneous estimation of total phenolic and alkaloid contents in the tea samples by utilizing the catechin and caffeine oxidation signals through the square-wave voltammetry technique. Food Chem 441. https://doi.org/10.1016/j.foodchem.2023.138262

139. Sadeghi A, Rajabiyan A, Nabizade N, Meygoli Nezhad N, Zarei-Ahmady A (2024) Seaweed-derived phenolic compounds as diverse bioactive molecules: a review on identification, application, extraction and purification strategies. Int J Biol Macromol 266
140. Sam AG, Song D (2022) ISO 14001 certification and industrial decarbonization: an empirical study. J Environ Manage 323. https://doi.org/10.1016/j.jenvman.2022.116169
141. Sampaio PDTB, Barbosa AP, Vieira G, Spironello WR, Bruno FMS (2005) Biomassa da rebrota de copas de pau-rosa (Aniba rosaeodora Ducke) em plantios sob sombra parcial em floresta primária. Acta Amaz 35:491–494. https://doi.org/10.1590/s0044-59672005000400014
142. Santos BLP, Vieira IMM, Ruzene DS, Silva DP (2024a) Unlocking the potential of biosurfactants: production, applications, market challenges, and opportunities for agro-industrial waste valorization. Environ Res 244
143. Santos E da N, Santana Neto DC de, Cordeiro ÂMT de M, Meireles BRL de A, Ferreira VC da S, Silva FAP da (2023) From waste to wonder: Unleashing the antimicrobial and antioxidant potential of acerola residue using a central composite rotatable design. J Environ Chem Eng 11(6):111184. https://doi.org/10.1016/J.JECE.2023.111184
144. Sar T, Harirchi S, Ramezani M, Bulkan G, Akbas MY, Pandey A, Taherzadeh MJ (2022) Potential utilization of dairy industries by-products and wastes through microbial processes: a critical review. Sci Total Environ 810:152253. https://doi.org/10.1016/j.scitotenv.2021.152253
145. Saravanan A, Karishma S, Senthil Kumar P, Rangasamy G (2023) A review on regeneration of biowaste into bio-products and bioenergy: life cycle assessment and circular economy. Fuel 338:127221. https://doi.org/10.1016/J.FUEL.2022.127221
146. Sarker A, Ahmmed R, Ahsan SM, Rana J, Ghosh MK, Nandi R (2023) A comprehensive review of food waste valorization for the sustainable management of global food waste. Sustain Food Technol 2:48–69
147. Sato MK, De Lima HV, Noronha Costa A, Rodrigues S, Mooney SJ, Clarke M, Silva Pedroso AJ, De Freitas Maia CMB (2020) Biochar as a sustainable alternative to açaí waste disposal in Amazon, Brazil. Process Saf Environ Prot 139:36–46. https://doi.org/10.1016/j.psep.2020.04.001
148. Scarpa ES, Antonelli A, Balercia G, Sabatelli S, Maggi F, Caprioli G, Giacchetti G, Micucci M (2024) Antioxidant, anti-inflammatory, anti-diabetic, and pro-osteogenic activities of polyphenols for the treatment of two different chronic diseases: type 2 diabetes mellitus and osteoporosis. Biomolecules 14
149. Schiebel CS, Bueno LR, Pargas RB, de Mello Braga LLV, da Silva KS, Fernandes ACVU, dos Santos Maia MH, de Oliveira NMT, Bach C, Maria-Ferreira D (2024) Exploring the biological activities and potential therapeutic applications of agro-industrial waste products through non-clinical studies: a systematic review. Sci Total Environ 950:175317. https://doi.org/10.1016/J.SCITOTENV.2024.175317
150. Seah CC, Habib SH, Hafriz RSRM, Shamsuddin AH, Salmiaton A (2023) Can waste eggshell replace commercial zeolites as catalyst for bio-oil production? J Anal Appl Pyrol 175:106213. https://doi.org/10.1016/j.jaap.2023.106213
151. Sefeedpari P, Pudełko R, Jędrejek A, Kozak M, Borzęcka M (2020) To what extent is manure produced, distributed, and potentially available for bioenergy? A step toward stimulating circular bio-economy in Poland. Energies 13:6266. https://doi.org/10.3390/en13236266
152. Sharma P, Vishvakarma R, Gautam K, Vimal A, Kumar Gaur V, Farooqui A, Varjani S, Younis K (2022) Valorization of citrus peel waste for the sustainable production of value-added products. Biores Technol 351:127064. https://doi.org/10.1016/j.biortech.2022.127064
153. Shinde R, Shahi DK, Mahapatra P, Singh CS, Naik SK, Thombare N, Singh AK (2022) Management of crop residues with special reference to the on-farm utilization methods: a review. Ind Crop Prod 181:114772. https://doi.org/10.1016/j.indcrop.2022.114772
154. Si Y, Luo H, Zhou F, Bai X, Han L, Sun H, Cha R (2021) Advances in polysaccharide nanocrystals as pharmaceutical excipients. Carbohydr Polym 262

155. Silva A, Rodrigues C, Garcia-Oliveira P, Lourenço-Lopes C, Silva SA, Garcia-Perez P, Carvalho AP, Domingues VF, Barroso MF, Delerue-Matos C, Simal-Gandara J, Prieto MA (2021) Screening of bioactive properties in brown algae from the northwest Iberian peninsula. Foods 10(8). https://doi.org/10.3390/foods10081915
156. Silva DW, Batista FG, Scatolino MV, Mascarenhas ARP, De Medeiros DT, Tonoli GHD, Lazo DAÁ, Caselli FDTR, De Souza TM, Alves Junior FT (2023) Developing a biodegradable film for packaging with lignocellulosic materials from the Amazonian biodiversity. Polymers 15:3646. https://doi.org/10.3390/polym15173646
157. Silva ICS, Da Silva PJ, De Carvalho FA, Da Silva SC, Yamashita F, Mendes ALD, Goto CA, Bosco TCD, Balan GC, Shirai MA (2024) Malt bagasse in extruded wheat flour and PBAT biodegradable films. Ind Crop Prod 220:119274. https://doi.org/10.1016/j.indcrop.2024.119274
158. Silveira Neto AL, Pimentel-Almeida W, Niero G, Wanderlind EH, Radetski CM, Almerindo GI (2023) Application of a biochar produced from malt bagasse as a residue of brewery industry in fixed-bed column adsorption of paracetamol. Chem Eng Res Des 194:779–786. https://doi.org/10.1016/j.cherd.2023.05.009
159. Singh M, Bhutani S, Dinkar N, Suyal DC, Perveen K, Bukhari NA, Khanam MN (2024) Assessment of pharmacological activities of specialized metabolites of Withania somnifera (L.). S Afr J Bot 166:259–271. https://doi.org/10.1016/J.SAJB.2024.01.039
160. Singh R, Das R, Sangwan S, Rohatgi B, Khanam R, Peera SKPG, Das S, Lyngdoh YA, Langyan S, Shukla A, Shrivastava M, Misra S (2021) Utilisation of agro-industrial waste for sustainable green production: a review. Environ Sustain 4:619–636. https://doi.org/10.1007/s42398-021-00200-x
161. Singhal S, Hulle NRS (2022) Citrus pectins: structural properties, extraction methods, modifications and applications in food systems—a review. Appl Food Res 2:100215. https://doi.org/10.1016/j.afres.2022.100215
162. Sivagnanam SP, Yin S, Choi JH, Park YB, Woo HC, Chun BS (2015) Biological properties of fucoxanthin in oil recovered from two brown seaweeds using supercritical CO2 extraction. Mar Drugs 13(6). https://doi.org/10.3390/md13063422
163. Skendi A, Zinoviadou KG, Papageorgiou M, Rocha JM (2020) Advances on the valorisation and functionalization of by-products and wastes from cereal-based processing industry. Foods 9:1243. https://doi.org/10.3390/foods9091243
164. Song Y, Maskey S, Lee YG, Lee D-S, Nguyen D-T, Bae H-J (2024) Optimizing bioconversion processes of rice husk into value-added products: D-psicose, bioethanol, and lactic acid. Biores Technol 395:130363. https://doi.org/10.1016/j.biortech.2024.130363
165. Srivastava N, Singh A, Kumari P, Nishad JH, Gautam VS, Yadav M, Bharti R, Kumar D, Kharwar RN (2020) Advances in extraction technologies: isolation and purification of bioactive compounds from biological materials. In: Natural bioactive compounds: technological advancements. Elsevier, pp 409–433
166. Tavares CS, Martins A, Miguel MG, Carvalheiro F, Duarte LC, Gameiro JA, Figueiredo AC, Roseiro LB (2020) Bioproducts from forest biomass II. Bioactive compounds from the steam-distillation by-products of Cupressus lusitanica Mill. and Cistus ladanifer L. wastes. Ind Crops Prod 158. https://doi.org/10.1016/j.indcrop.2020.112991
167. Tijjani H, Zangoma MH, Mohammed ZS, Obidola SM, Egbuna C, Abdulai SI (2020) Polyphenols: classifications, biosynthesis and bioactivities. In: Egbuna C, Dable Tupas G (eds) Functional foods and nutraceuticals. Springer International Publishing, Cham, pp 389–414
168. Tomar GS, Gundogan R, Can Karaca A, Nickerson M (2023) Valorization of wastes and by-products of nuts, seeds, cereals and legumes processing. In: Advances in food and nutrition research. Academic Press Inc., pp 131–174
169. Tripathi N, Hills CD, Singh RS, Atkinson CJ (2019) Biomass waste utilisation in low-carbon products: harnessing a major potential resource. Clim Atmos Sci 2:35. https://doi.org/10.1038/s41612-019-0093-5
170. U.S Department of Agriculture—USDA (2024) Agricultural projections to 2033. https://www.usda.gov/sites/default/files/documents/USDA-Agricultural-Projections-to-2033.pdf. Accessed 16 Sept 2024

171. U.S Department of Agriculture—USDA (2024) Citrus: world markets and trade. https://fas.usda.gov/sites/default/files/2024-07/citrus.pdf. Accessed 16 Sept 2024
172. Ucella-Filho JGM, Ferreira NS, Cupertino GFM, Guedes NA, de Paula H, Kitagawa RR, Balu AM, Luque R, Ignacchiti MDC, Parreira LA, Resende JA, Dias Junior AF (2024) Antiviral potential of bio-oil from Citrus sinensis waste wood as a therapeutic approach against COVID-19. J Clean Prod 447:141583. https://doi.org/10.1016/J.JCLEPRO.2024.141583
173. Ucella-Filho JGM, Lorenço MS, da Silva BRF, Ferreira VRF, Monteiro JRB, Ferreira NS, das Cardoso MG, Kitagawa RR, Resende JA, Dias Junior AF, Piccoli RH, Mori FA (2024) Exploring the potential of tannin-rich tree bark extracts in combating foodborne diseases and gastric cancer. Food Biosci 57:103559. https://doi.org/10.1016/J.FBIO.2023.103559
174. Ucella-Filho JGM, Ferreira NS, Lacerda EMF, Soriano Neto C, Duarte CEDS, Ferreira IM, Balu AM, Ignacchiti MDC, Dias Junior AF, Resende JA (2024) Citrus sinensis waste wood pyroligneous extract as an effective agent against biofilms formed by mastitis-causing Staphylococcus aureus. Ind Crops Prod 218:118854. https://doi.org/10.1016/j.indcrop.2024.118854
175. United Nations—UN (2023) Department of Economic and Social Affairs, Population Division. India overtakes China as the world's most populous country. https://www.un.org/development/desa/dpad/publication/un-desa-policy-brief-no-153-india-overtakes-china-as-the-worlds-most-populous-country/. Accessed 18 Sept 2024
176. United Nations Environment Programme–UNEP (2009) Converting waste agricultural biomass into a resource. United Nations Environment Programme Division of Technology, Industry and Economics International Environmental Technology Centre, Osaka/Shiga, Japan
177. Venkatramanan V, Shah S, Rai AK, Prasad R (2021) Nexus between crop residue burning, bioeconomy and sustainable development goals over North-Western India. Front Energy Res 8:614212. https://doi.org/10.3389/fenrg.2020.614212
178. Villacreces S, Blanco CA, Caballero I (2022) Developments and characteristics of craft beer production processes. Food Biosci 45:101495. https://doi.org/10.1016/j.fbio.2021.101495
179. Visakh NU, Pathrose B, Chellappan M, Ranjith MT, Sindhu PV, Mathew D (2023) Extraction and chemical characterisation of agro-waste from turmeric leaves as a source of bioactive essential oils with insecticidal and antioxidant activities. Waste Manage 169:1–10. https://doi.org/10.1016/J.WASMAN.2023.06.030
180. Wang Y, Wang Z, Lin Y, Qin Y, He R, Wang M, Sun Q, Peng Y (2024) Nanocellulose from agro-industrial wastes: a review on sources, production, applications, and current challenges. Food Res Int 192
181. Yadav P, Singh S, Jaiswal S, Kumar R (2024) Synthetic and natural polymer hydrogels: a review of 3D spheroids and drug delivery. Int J Biol Macromol 280
182. Yadav S, Malik K, Moore JMC, Kamboj BR, Malik S, Malik VK, Arya S, Singh K, Mahanta S, Bishnoi DK (2024) Valorisation of agri-food waste for bioactive compounds: recent trends and future sustainable challenges. Molecules 29
183. Yu Y, Shen M, Song Q, Xie J (2018) Biological activities and pharmaceutical applications of polysaccharide from natural resources: a review. Carbohydr Polym 183:91–101
184. Yun H, Hwang S-W, Jung M, Choi I-G, Yeo H, Kwak HW (2024) Microwave-assisted utilization of kraft lignin-derived activated carbon for efficient dye removal. Biomass Bioenerg 186:107279. https://doi.org/10.1016/j.biombioe.2024.107279
185. Zhang Q, Zhou Y, Yue W, Qin W, Dong H, Vasanthan T (2021) Nanostructures of protein-polysaccharide complexes or conjugates for encapsulation of bioactive compounds. Trends Food Sci Technol 109:169–196. https://doi.org/10.1016/J.TIFS.2021.01.026
186. Zhang Y, Hawboldt K, MacQuarrie S (2024) Extraction of bioactive compounds from beach-cast brown algae: a review on accelerated solvent extraction and subcritical water extraction. RSC Sustain 2:2069–2091

Potential Use of Pyroligneous Liquid Derived from Biomass as a Bioactive Product for Pharmaceutical Development

João Gilberto Meza Ucella-Filho, Gabriela Fontes Mayrinck Cupertino, Fabíola Martins Delatorre, Matheus Fernandes de Carvalho Reis, Iara Fontes Demuner, and Allana Katiussya Silva Pereira

Abstract Pyroligneous liquid is a condensate obtained from the gases released during the biomass pyrolysis process, mainly composed of water and a complex mixture of organic compounds, such as carboxylic acids, aldehydes, ketones, phenols, and hydrocarbons. Its properties vary depending on the biomass used and the specific conditions of the pyrolysis process, such as the heating rate and final temperature. This material has attracted significant interest from both the scientific community and industry due to its diverse applications, including the development of new pharmaceutical products. The chemical complexity of pyroligneous liquids is responsible for their various bioactive actions, such as antimicrobial, antioxidant, antiviral, cytotoxic, and anti-inflammatory activities. However, studies in this area are still relatively recent. Therefore, in this chapter, we will present a bibliometric analysis, highlighting the main trends, advancements, and the evolution of research on the potential use of pyroligneous liquid in disease treatment. Next, we will explore pyroligneous liquid

J. G. M. Ucella-Filho (✉) · M. F. de Carvalho Reis · I. F. Demuner
Department of Forestry Engineering, Universidade Federal de Viçosa (UFV), Viçosa, MG 36570-900, Brazil
e-mail: joao.meza@ufv.br

M. F. de Carvalho Reis
e-mail: matheus.reis@ufv.br

I. F. Demuner
e-mail: iara.demuner@ufv.br

G. F. M. Cupertino · F. M. Delatorre
Agricultural Sciences and Engineering Center, Universidade Federal do Espírito Santo (Ufes), Av. Governador Lindemberg, 316, Jerônimo Monteiro, ES 29550-000, Brazil
e-mail: gabriela.mayrinck01@gmail.com

F. M. Delatorre
e-mail: fabiolamdelatorre@gmail.com

A. K. S. Pereira
Department of Forest Sciences, "Luiz de Queiroz" College of Agriculture (Esalq), Universidade de São Paulo (USP), Av. Pádua Dias, 11, Piracicaba, SP 13418-900, Brazil
e-mail: allana.florestal@gmail.com

A. K. Silva Pereira and A. F. Dias Júnior (eds.), *Agro-industrial Waste for New Pharmaceuticals*, Green Energy and Technology,
https://doi.org/10.1007/978-3-031-88534-1_2

derived from forest and agricultural biomass as a bioactive product, focusing on its potential for the treatment of infectious and chronic diseases, consolidating the state of the art of its pharmaceutical applications.

Keywords Forest biomass · Agricultural biomass · Liquid smoke · Bio-oil · Natural products

1 Introduction

Climate change is one of the main global threats of the twenty-first century, driven by greenhouse gas (GHG) emissions resulting from fossil fuel combustion, deforestation, and unsustainable agricultural practices. These factors are continuously raising average global temperatures, altering precipitation patterns, and intensifying extreme weather events, such as hurricanes and heatwaves [17, 81]. Projections indicate that by the end of the century, average global temperatures could increase by 1.4 to 5.8 °C, accompanied by rising sea levels [124]. In a moderate scenario, exposure to coastal storm flooding could rise from 75 to 200 million people by the 2080s, with an expected 40 cm rise in sea level [101]. There is growing evidence that these changes are already impacting human health, increasing mortality and morbidity associated with extreme heat, severe cold, droughts, and storms [17, 81, 156]. Rising temperatures and humidity promote the proliferation of disease vectors, such as mosquitoes, which transmit viruses responsible for diseases like dengue and malaria [136]. Simultaneously, extreme weather events affect water and food quality, facilitating the spread of pathogenic bacteria and fungi that cause severe infections.

The SARS-CoV-2 virus, known as Coronavirus (COVID-19), clearly illustrates the impact of climate change on global health. Since its emergence, the virus has rapidly spread, resulting in a pandemic that, by April 2024, caused over 7 million deaths [155]. This event underscores the vulnerability of societies to health crises exacerbated by climate change, highlighting the urgent need for effective preventive measures to protect public health and mitigate the devastating effects of such emergencies. The response to COVID-19 involved the development and implementation of various medications and treatments, which played a crucial role in managing the pandemic. However, the production and distribution of these treatments faced significant challenges, including resource limitations and inequalities in access, exacerbated by high global demand [197]. These challenges are not exclusive to viruses but apply to any pathogen causing infectious diseases. The demand for rapid and accessible therapies has underscored the importance of a robust and well-coordinated infrastructure to handle large-scale health emergencies. This context reinforces the importance of continuous investment in research, development, and the distribution of medications, as well as swift and efficient responses to safeguard public health during future crises.

Given the current global health challenges, reducing costs and expanding access to medications are fundamental to ensuring that therapeutic advancements reach all segments of the population. In this context, the search for new bioactive compounds for medications becomes a priority, aiming to address emerging diseases and chronic conditions in a more accessible and sustainable manner. Increasing resistance to conventional treatments, such as antibiotics and antivirals, intensifies the demand for effective and lower-cost therapeutic alternatives [33]. Bioactives derived from natural and renewable sources, such as plants, stand out for their diversity and potential to reduce production costs and environmental impact, promoting sustainability in the pharmaceutical industry [103]. Plants are rich sources of bioactive compounds for medications, including alkaloids, which possess significant pharmacological effects [4], flavonoids, known for their antioxidant and anti-inflammatory properties [117], phenolic compounds, which have strong antioxidant activity [137], terpenoids, valued for their anti-inflammatory, antiviral, and anticancer properties [2, 150], and tannins, used for their antimicrobial and astringent properties [173, 174]. The diversity of plant compounds is crucial for developing new treatments, and continuous research is vital for creating effective and accessible medications.

The growing demand for natural and sustainable sources of bioactive compounds for the development of new medications has led to the enhancement of various techniques, among which pyrolysis stands out. This thermal process decomposes organic materials in conditions of absence or limited oxygen, resulting in solid, liquid, and gaseous products [37]. Pyrolysis utilizes a variety of biomass, including agricultural, forestry, and urban materials, transforming waste into bio-oil, charcoal, and syngas [66, 169, 176]. This conversion not only minimizes waste but also adds value to materials that would otherwise be discarded, such as crop residues, sugarcane bagasse, and food processing byproducts. Thus, pyrolysis emerges as a promising solution for waste valorization and research into the development of new pharmaceutical products, particularly due to its pyroligneous liquid, which has gained prominence in the context of bioactives [41, 171]. Pyroligneous liquid consists of an organic and an aqueous fraction, both containing a complex mixture of high-value compounds [76]. This diverse composition gives pyroligneous liquid significant potential for applications in various areas, including therapeutic treatments. Among these compounds are alcohols, acetones, esters, phenols, hydrocarbons, and nitrogenous compounds [36].

The composition and concentration of these substances in pyroligneous liquid can be heavily influenced by variables such as pyrolysis temperature, heating rate, type of biomass, and storage conditions [32], Ucella-Filho et al. [171]. Each of these conditions uniquely impacts the chemical structure of bio-oil, modifying both the composition and concentration of bioactive substances [60]. Among the most determining factors, pyrolysis temperature stands out as it directly influences the thermal decomposition of biomass and product distribution. At lower temperatures, pyroligneous liquid presents a higher concentration of oxygenated compounds, such as phenols, alcohols, and ketones. Conversely, at higher temperatures, the formation of aromatic hydrocarbons and other volatile compounds increases [27]. This variation in composition impacts the presence of bioactive compounds with antioxidant, antimicrobial, and anti-inflammatory properties, which are crucial for drug development. The heating rate also plays a significant role in the quality of pyroligneous liquid. A slower heating rate favors the formation of phenolic compounds, resulting in a pyroligneous liquid richer in these compounds [174]. Phenols are known for their antioxidant, anti-inflammatory, antimicrobial, and anticancer activities, making them highly promising for the pharmaceutical industry [171].

Therefore, precise control of parameters such as temperature and heating rate can optimize the production of specific compounds, expanding the potential of pyroligneous liquid as a promising source of bioactives for the development of new drugs. Additionally, the type of biomass used in the pyrolysis process significantly influences the chemical profile of bio-oil [106]. Variations in the chemical composition of bio-oil are fundamental for its application in pharmaceutical contexts, as certain bioactive compounds possess specific properties that can be leveraged in drug formulation. Research interest in bio-oil derived from agricultural residues has grown, with studies highlighting its potential as a therapeutic agent. Utilizing these sustainable sources not only benefits public health but also adds value to agricultural residues, contributing to circular economy practices and sustainability. Thus, the objective of this work is to present a review of recent research investigating the use of pyroligneous liquid as a source of bioactive compounds, focusing on the differences between forest and agricultural biomasses used in the pyrolysis process.

2 State of the Art: Use of Pyroligneous Liquid as a Bioactive Product

Bibliometric analysis is a tool used to quantify and visualize the scientific development of a specific research niche. In this context, we employed bibliometric analysis to understand the current state of research evaluating the use of pyroligneous liquid obtained from different forest and agricultural biomasses as a bioactive product for the treatment of infectious and chronic diseases. To achieve this, we selected the Scopus database, recognized as one of the largest scientific repositories and suitable for bibliometric analyses, following the methodology defined by Ucella-Filho et al. [170]. To conduct the survey, we defined keywords that would guide the search toward studies aligned with the objective of this chapter. The survey was performed using the following query: (TITLE-ABS-KEY ("bio-oil" OR "pyroligneous acid" OR "pyroligneous liquid" OR "wood vinegar" OR "pyroligneous extract") AND TITLE-ABS-KEY ("antimicrobial" OR "antibacterial" OR "antifungal" OR "antiviral" OR "antitumor" OR "antioxidant" OR "cytotoxic" OR "antiinflammatory" OR "anticancer")) AND (LIMIT-TO (DOCTYPE, "ar")). For this analysis, review articles, book chapters, and books were excluded, retaining only research articles. The collected data were processed using the "biblioshiny" function in the "bibliometrix" package [7] from R Core Team software (2021).

A total of 316 documents were identified, of which 256 corresponded to research articles and were included in this data analysis. The first article published on the bioactive potential of pyroligneous liquid dates back to 1997, titled Evaluación Antimicrobial de 4 productos derivados del ácido piroleñoso para su posible utilización como desinfectantes, published in *Revista Cubana de Farmácia* by Miriam Concepción Rojas and collaborators. Publications began to grow steadily after 2007, with 2022 being the year with the highest number of publications, totaling 42 documents (Fig. 1). This significant increase in publications in 2022 may be associated with the COVID-19 pandemic period, during which researchers, due to isolation and travel restrictions (e.g., access to research laboratories) [65], focused on manuscript writing and submission for publication. Furthermore, pyroligneous liquid, due to its proven bioactive properties [41], became the subject of studies evaluating its efficacy in treating various diseases caused by bacteria, fungi, and viruses. However, it was only in 2024 that an article evaluated the potential of pyroligneous liquid in the treatment of COVID-19. This study, developed by João Gilberto Meza Ucella-Filho and collaborators, was published in the *Journal of Cleaner Production* under the title "Antiviral potential of bio-oil from Citrus sinensis waste wood as a therapeutic approach against COVID-19".

The evaluation of pyroligneous liquid as a bioactive product is a growing research field across different regions of the world. The survey revealed that 47 countries have conducted studies in this area, with China, Brazil, Malaysia, and the United States being the top contributors, publishing 56, 30, 18, and 18 articles, respectively. Pyroligneous liquid is obtained during the pyrolysis process, which involves the thermochemical transformation of biomass. This process generates not only pyroligneous liquid but also charcoal and non-condensable gases, with charcoal being the primary product of pyrolysis. In this context, the high production of charcoal in

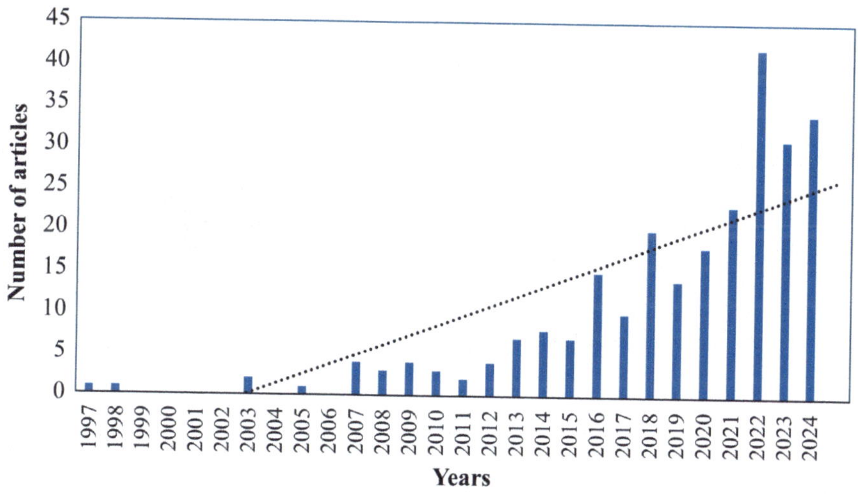

Fig. 1 Scientific production over the years. *Source* Authors (2024)

China and Brazil [72] may explain the elevated number of articles published, as one way to mitigate the impact of gases released during carbonization is to condense these gases, resulting in the formation of pyroligneous liquid. The use of residues for various purposes has driven research efforts aimed at adding value to these byproducts. Due to its chemical composition, pyroligneous liquid has significant potential as the basis for developing various products, including medications for treating different diseases.

Figure 2 illustrates the collaborations between countries in the development of these studies; the thicker the line connecting regions, the greater the number of published articles. A total of 63 collaborations were identified, with notable partnerships between China and the United States, resulting in 5 publications, followed by Brazil and Spain, with 3 publications. International cooperation is crucial for strengthening relationships between researchers and institutions, fostering the production of high-impact scientific work and promoting collaborative solutions to global challenges [47, 56]. Additionally, collaboration between countries enables scientific progress by facilitating information exchange, fostering the emergence of new ideas for innovative research development. Conversely, the lack of knowledge exchange can lead to similar research goals being pursued independently in different regions, limiting innovation and scientific advancement.

A total of 762 keywords were defined by the authors of the articles. Figure 3 presents the trend topics with the 50 most frequently used keywords. The most common terms include wood vinegar (n = 68), pyroligneous acid (n = 50), bio-oil (n = 41), pyrolysis (n = 41), antioxidant activity (n = 22), and antifungal activity (n = 17). These keywords indicate the focus of the research, highlighting the greatest interest in this study niche.

From the analysis, it is evident that there is a lack of standardization in the terms used to refer to the condensable gases generated during pyrolysis. Authors use various denominations such as "bio-oil", "wood vinegar", "pyroligneous acid', "pyroligneous liquid', among others, to describe the same product. This lack of uniformity

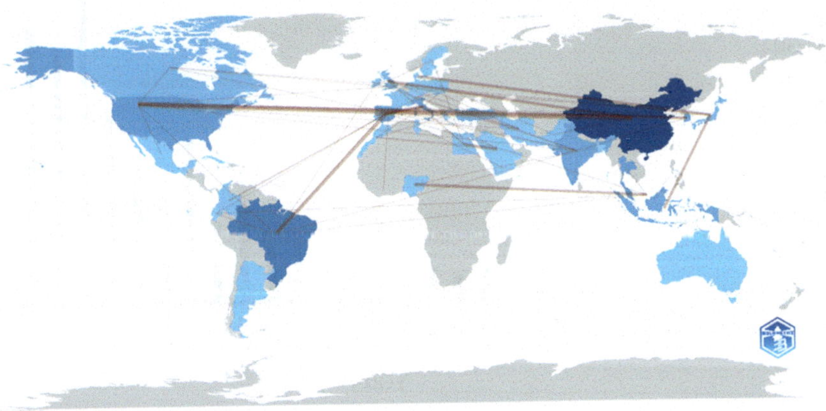

Fig. 2 Collaborative network between countries. *Source* Authors (2024)

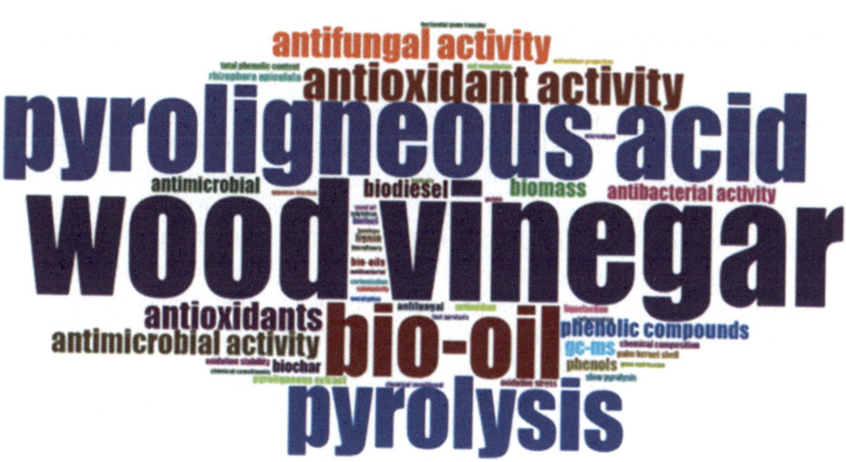

Fig. 3 Main keywords used by the authors

can pose significant challenges in searching for and categorizing articles, potentially underestimating the volume of existing research on the topic. Moreover, this diversity in nomenclature may hinder the comparison of results between different studies, complicating the consolidation of knowledge in the field. To address these issues, it is essential for researchers in this area to work towards a more unified terminology.

The keywords also reveal that the use of pyroligneous liquid as a bioactive product focuses mainly on its antioxidant and antifungal properties. Oxidative stress is associated with a variety of diseases, including cancer, neurodegenerative disorders, and cardiovascular conditions [131], making the evaluation of the antioxidant potential of pyroligneous liquid essential. Bioactive compounds present in pyroligneous liquid can neutralize free radicals, providing cellular protection against oxidative damage, thus contributing to disease prevention. Furthermore, the antifungal activity of pyroligneous liquid becomes increasingly relevant in the context of rising microbial resistance and the search for natural alternatives to synthetic antifungals [55]. As resistance to conventional drugs becomes a growing challenge in medicine, pyroligneous liquid emerges as a promising option, with the potential to develop new therapies effective against resistant pathogens.

The bibliometric analysis highlighted the main challenges and gaps present in this research field and served as the foundation for structuring the subsequent topics of this chapter. More detailed information about the use of pyroligneous liquid derived from forest and agricultural biomasses will be presented in the following sections, emphasizing the critical aspects that should be considered for the development of innovative research utilizing pyroligneous liquid as a bioactive product.

3 Thermochemical Conversion of Biomass

Pyrolysis is one of the most promising technologies for converting agricultural biomass into value-added products such as pyroligneous liquid, gases, and biochar [111]. The thermal decomposition of biomass occurs in an environment with limited or no oxygen, preventing combustion and allowing the breakdown of biomass components into smaller chemical substances [106]. Among thermochemical processes such as combustion and gasification, pyrolysis stands out for its ability to operate under controlled conditions, maximizing product efficiency. Unlike combustion, which generates energy through burning, and gasification, which produces combustible gases, pyrolysis focuses on the controlled decomposition of biomass, preserving the characteristics of its products, especially the liquid fraction—pyroligneous liquid—which holds significant value [34]. This approach allows for optimizing yield and adjusting product quality, with an emphasis on pyroligneous liquid, expanding its application possibilities in energy, chemical, and pharmaceutical sectors. Pyroligneous liquid is a complex fraction composed of a wide variety of compounds, such as acids, ketones, aldehydes, phenols, alcohols, and hydrocarbons [76]. Its composition can vary significantly depending on the operational parameters of the pyrolysis process [121, 158, 167, 188]. This approach allows for the optimization of yield and adjustment of product quality, with an emphasis on pyroligneous liquid, thereby expanding its potential applications in the energy, chemical, and pharmaceutical sectors [36, 174].

However, the quality of pyroligneous liquid depends directly on factors such as reaction temperature, heating rate, vapor residence time, and biomass characteristics [73, 147]. Temperature is one of the main factors influencing biomass decomposition and the quality of products obtained during pyrolysis [48]. At lower temperatures (generally up to 450 °C), the yield of pyroligneous liquid tends to be lower, but the liquid fraction contains compounds with a high oxygen content, such as acids and alcohols, which compromise its stability and energy potential [109]. As the temperature increases (between 500 °C and 800 °C), long-chain organic compounds break down, resulting in compounds with lower oxygen content and, therefore, greater stability [48]. The temperature effect also depends on the biomass components, such as cellulose, hemicellulose, and lignin. For instance, lignin decomposes at higher temperatures, generating aromatic compounds, while cellulose and hemicellulose decompose at moderate temperatures, producing liquids rich in sugars and oxygenated compounds [183]. Thus, selecting the operating temperature is crucial to balancing the yield and quality of pyroligneous liquid, tailoring it to the desired application.

The heating rate plays a crucial role in the efficiency of pyrolysis, directly influencing product formation. Rapid heating rates favor the immediate release of volatiles, promoting the formation of pyroligneous liquid [99]. The quick evaporation of compounds prevents secondary reactions, such as polymerization or cracking, which could lead to the formation of undesirable gases or solids [69]. As a result, a lighter pyroligneous liquid with lower tar content is obtained, increasing its quality and application potential. On the other hand, slower heating rates allow for greater interaction of vapors within the reaction zone, favoring the formation of more complex and difficult-to-condense compounds [135]. This can result in lower liquid yields and higher production of biochar and gas. Therefore, selecting the appropriate heating rate is essential to optimizing pyroligneous liquid production, maximizing both yield and product quality, especially when the focus is on obtaining bioactive compounds [174]. Slow heating rates can better preserve phenolic and antioxidant compounds, which have valuable applications in the pharmaceutical and cosmetics industries [174]. Controlled heating rates also influence the formation of complex molecules with antimicrobial and anti-inflammatory properties, expanding the potential of pyroligneous liquid as a renewable source of bioactive products with commercial interest.

In addition to operational parameters, the chemical composition of agricultural residues used in pyrolysis is a determining factor in the quality of pyroligneous liquid. Agricultural residues comprise a variety of materials, including cellulose, hemicellulose, lignin, and other organic and inorganic components [110]. This chemical diversity directly impacts thermal decomposition during pyrolysis, influencing both the yield and the composition of generated products. Understanding how these characteristics affect pyroligneous liquid quality is crucial for developing specific applications. By analyzing residue composition and its reactions during pyrolysis, processes can be optimized to produce liquids that meet specific demands in the energy and pharmaceutical sectors. This approach not only maximizes the value of agricultural and forest biomass but also contributes to sustainability by transforming byproducts into valuable resources.

4 Main Methods of Thermochemical Conversion of Biomass

The primary methods for thermochemical conversion of wood and forest biomass include combustion, pyrolysis, and gasification. Biomass combustion is mainly used for heat and energy generation [157]. Pyrolysis and gasification processes, on the other hand, are capable of converting wood and/or biomass into heat, solid, liquid, and gaseous products [180, 190], making biomass an affordable, clean, and environmentally friendly energy source [148, 149]. Gasification primarily produces gaseous products, with a typical yield of 85% gas, 10% solids, and 5% liquids. Slow pyrolysis has a more balanced product distribution, yielding 35% gas, 35% solids, and 30% liquids. Fast and flash pyrolysis, in contrast, typically yield 75% liquids, 13% gases, and 12% solids [61]. Figure 4 presents the summary of yields for methods of thermochemical conversion of biomass. As a result, pyrolysis is the most widely used process for converting biomass into liquids. Pyrolysis is a thermal decomposition process of organic compounds in the absence of oxygen. During this process, the cellulose, hemicellulose, and lignin components of biomass break down [84]. Pyrolysis can be divided into four stages: (1) Initial decomposition of biomass (<215 °C), (2) Decomposition of hemicellulose (215–375 °C), (3) Decomposition of cellulose and lignin (375–500 °C) and (4) Decomposition of lignin (>500 °C) [85]. These stages involve breaking the chemical bonds of biomass components, releasing gases (condensable and non-condensable), and rearranging reaction products into a residue matrix [35].

The pyrolysis process has been extensively developed to produce fuels and chemicals from various types of biomasses [182], including forest biomass. Utilizing coproducts derived from pyrolysis is essential for optimizing forest resources [13]. The solid products from pyrolysis are known as biochar, the gaseous components as pyrolysis gases, and the liquid fraction as pyroligneous liquid [61]. Four types of pyrolysis processes can be adopted: slow, intermediate, fast, and flash. Slow pyrolysis: Also known as carbonization, has been used for product generation for over a thousand years [182]. This method involves a slow heating rate (0.1–1 °C/s), favoring charcoal production [61, 70]. Slow pyrolysis also requires moderate temperatures (400–500 °C) and longer residence times, measured in minutes [61]. Intermediate pyrolysis: Typically operates at similar temperatures (400–500 °C), with slow heating rates and intermediate residence times (~5–10 min) [86]. Fast pyrolysis: Emerged

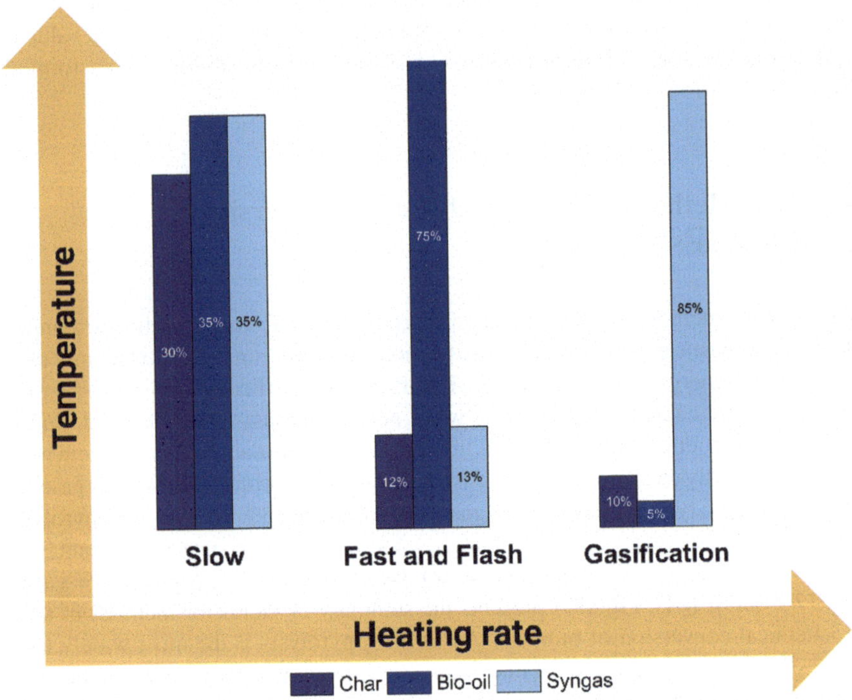

Fig. 4 Yields for methods of thermochemical conversion of biomass. *Source* Authors (2024)

in the 1980s, driven by the oil crisis [182], as a means of increasing pyroligneous liquid yield. Technologies such as ablative reactors, rotating cone reactors, bubbling fluidized beds, circulating fluidized beds, and entrained flow beds were developed to facilitate the process [19, 182]. Fast pyrolysis operates at higher heating rates (10–200 °C/s), elevated temperatures (450–650 °C), and shorter residence times (2–3 s) [49, 61]. Flash pyrolysis: Aims to maximize liquid and gas yield over charcoal. It operates at higher temperatures (650–1300 °C), faster heating rates (100–1000 °C/s), and shorter residence times (<1 s) compared to rapid pyrolysis [70, 86].

5 Forestry Sector and Potential of Forest Biomass

Forests cover more than one-third of the Earth's surface and provide essential ecological services [154]. These include climate regulation, erosion control, water quality improvement, soil development, nutrient cycling, air quality maintenance, and biodiversity preservation [75, 113, 145]. Additionally, global forests absorb approximately 15.6 billion tons of CO_2 annually [63, 79], contributing significantly to carbon sequestration. The carbon stored in forests is predominantly found in biomass such as roots, trunks, and branches [79].

Sustainability practices are implemented by forestry companies, involving investments in conservation, local community support, and mitigation of environmental impacts caused by production processes [179]. Consequently, the forestry sector makes a significant contribution to climate action, including reducing greenhouse gas emissions (GHG) and global warming [198]. Forests are also utilized for product generation [113, 154], with examples including cellulose, paper, charcoal, sawn timber, and engineered wood panels.

Primary forest biomass refers to organic material directly derived from living or recently dead trees, as well as byproducts from wood harvesting. Secondary forest biomass is generated by forestry industries, including wood chips, sawdust, and black liquor [96]. Eucalyptus and pine woods are notable for biomass production and conversion into products. Forest wood and biomass are primarily composed of cellulose, hemicellulose, and lignin, with quantitative proportions varying according to the raw material. Generally, cellulose is the main component of cell walls, representing 40–60% by weight, followed by hemicellulose (20–40%) and lignin (10–30%) [24, 97, 195]. Small amounts of extractives and ashes are also present in these raw materials [5].

Cellulose is the most abundant biopolymer in nature, consisting of D-glucopyranose units linked by β-1,4-glycosidic bonds, with highly ordered (crystalline) and disordered (amorphous) regions [59]. Hemicellulose chains may include five- and six-carbon sugars, acidic components (galacturonic and glucuronic acids), and acetyl groups, with variability depending on the species [90]. Unlike cellulose, hemicellulose is amorphous, with low molecular weight and polymerization degree [91]. Lignin is the second most abundant biopolymer in nature, surpassed

Table 1 Chemical composition of forest biomasses reported in scientific studies

Forest biomass	Components (wt%)			Reference
	Cellulose	Hemicellulose	Lignin	
Picea spp. wood	42	27	26	Burhenne et al. [20]
Picea spp. bark	48.10 ± 1.10	19.75 ± 0.53	22.87 ± 0.48	Armanu et al. [8]
Picea spp. sawdust	55.40 ± 1.3	5.6 ± 0.6	28.70 ± 1.0	Auxenfans et al. [12]
Pinus spp. sawdust	41.00 ± 0.36	18.89 ± 0.26	25.54 ± 1.67	Cavali et al. [23]
Eucalyptus spp. wood	35–56	9–23	21–37	Carrillo et al. [22]
Eucalyptus spp. sawdust	42.67 ± 0.80	16.05 ± 0.24	23.73 ± 0.10	Tavares et al. [164]
Citrus sinensis wood residue	40.50 ± 0.30	29.30 ± 1.80	21.20 ± 0.70	Rosado et al. [138]
Populus spp. wood	35.85	40.90	23.25	Boundzanga et al. [18]
Populus alba L. sawdust	55.78	14.94	20.20	Selvi Gökkaya et al. [146]
Quercus spp. wood	38–47	19–30	22–30	Le Floch et al. [82]
Quercus spp. sawdust	44.70 ± 0.3	16.00 ± 1.4	26.70 ± 0.8	Auxenfans et al. [12]
Hardwood bark	10–25	11–23	23–28	Millati et al. [104]
Softwood bark	19–25	11–18	23–28	Millati et al. [104]

only by cellulose [139]. It is a polymer composed of three phenylpropane units: p-coumaryl alcohol (H), coniferyl alcohol (G), and sinapyl alcohol (S), with proportions varying by species. Lignin has a complex, heterogeneous, aromatic, and polydisperse structure [74].

The chemical properties of forest biomass vary significantly, with differences observable between species, within species, and even within a single tree [6]. This variability enables the optimal use of each raw material for conversion into products, improving process efficiency and final product quality. Some chemical compositions of forest biomasses reported in the literature are summarized in Table 1.

5.1 Properties of Pyroligneous Liquid from Forest Biomass and Its Bioactive Properties

Pyroligneous liquid is a highly complex mixture of chemical components [182], formed through the simultaneous depolymerization and fragmentation of cellulose, hemicellulose, and lignin from forest biomass [196]. The liquid derived from forest biomass degradation comprises water and chemical compounds, including phenols, organic acids, ketones, aldehydes, alcohols, esters, furan, and pyran derivatives, among others [58, 153, 182, 184]. The chemical composition of pyroligneous liquid depends on the characteristics of the forest biomass and the conditions of the carbonization process [94]. Typically, the amount of phenolic compounds in pyroligneous liquid increases with higher pyrolysis temperatures [78]. Phenolic compounds and their derivatives are formed through lignin degradation, while organic acids, alcohols, aldehydes, and ketones originate from the thermal degradation of hemicellulose and cellulose.

The hemicellulose content in forest biomass may contribute to higher acidity in pyroligneous liquid due to the production of organic acids during its thermal degradation, primarily acetic acid. These acids result in a low pH [16, 19], which can cause pipe corrosion during application. Alkali metals and chloride, present in forest biomass, may remain as components of pyroligneous liquid, altering its quality [86]. Regarding the quality of forest-derived liquid and the pyrolysis process, rapid pyrolysis yields higher liquid production but results in lower calorific value due to the high content of H_2O, O_2, and carboxylic acids [61]. The high oxygen content leads to reduced energy density, stability, and miscibility with hydrocarbons in pyroligneous liquid [86]. However, water content reduces viscosity and increases fluidity, which is beneficial for atomization and combustion [196]. Physicochemical properties of pyroligneous liquid obtained from wood pyrolysis have been reported by Grewal et al. [58]: moisture content (15–30 wt%), pH (2.5), elemental composition of carbon (54–58%), hydrogen (5.5–7.0%), oxygen (35–40%), nitrogen (0–0.2%), and ash (0–0.2%), higher calorific value (16–19 MJ/kg), viscosity at 50 °C (40–100 cP), and total solids (0.2–1.0 wt%). Understanding the chemical composition and physical properties of pyroligneous liquid is crucial for determining its stability and suitability for industrial applications [19].

The use of pyroligneous liquid is increasingly gaining attention due to its unique properties and sustainable appeal. Its chemical composition and characteristics make it versatile for various applications, including agriculture, medicine, wood preservation, and food preservation [120]. In the pharmaceutical context, pyroligneous liquid has been studied for its bioactive properties, such as antioxidant, antimicrobial, and anti-inflammatory effects [125]. The main medical and pharmaceutical applications of interest found in the literature are summarized in Table 2.

Pyroligneous liquid is rich in phenolic compounds and has the potential to be used in treatments aimed at wound healing, promoting tissue regeneration and accelerating the healing process [125]. Moreover, its antioxidant and anti-inflammatory properties make it a promising ingredient in the pharmaceutical industry [108]. Pyroligneous liquid has also shown potential as a candidate for the development of new antibiotics for humans and animals due to its antimicrobial properties against pathogenic microorganisms [125].

In other industries, pyroligneous liquid is widely used as a natural pesticide and fertilizer in agriculture. It also serves as a biological pest controller while improving soil quality and promoting plant growth [120]. In the veterinary and animal production context, pyroligneous liquid can be added to animal feed as an antibiotic substitute, helping control pathogens and promoting animal health [125]. The use of pyroligneous liquid also extends to the food industry, where it is employed as a smoking agent, imparting flavor and being used in food preservation [89].

Table 2 Overview of the use of pyroligneous liquid in the pharmaceutical context and other applications reported in scientific studies

Author	Biomass	Type of conversion	Application
Ouattara et al. [119]	Wood mix of *Cola nitidae*; *Terminalia ivorensis*; *Havea brasiliensis*; *Acacia mangium*; *Petersianthus macrocarpus*; *Piptadeniastrum africanum*; *Milicia excelsa* and *Khaya ivorensis*	Slow pyrolysis	Antioxidant
Ucella-Filho et al. [172]	*Citrus sinensis* wood	Slow pyrolysis	Antibacterial
Oramahi et al. [118]	*Terminalia catappa* wood	Slow pyrolysis	Antifungal
Yu et al. [192]	*Larix gmelinii* (Rupr.) *Kuzen* wood	Fast pyrolysis	Antifungal
Chang et al. [25]	Hardwood	Slow pyrolysis	Antibacterial
Rizal et al. [134]	*Melaleuca Leucadendron* biomass	Slow pyrolysis	Antibacterial and Antioxidant
Ucella-Filho et al. [172]	*Citrus sinensis* wood	Slow pyrolysis	Antiviral
El-Fawy et al. [50]	*Psidium guajava* wood	Slow pyrolysis	Antifungal
Gama et al. [57]	*Eucalyptus urophylla* x *Eucalyptus grandis*, clone I144 wood	Slow pyrolysis	Antibacterial and Antifungal
Feng et al. [54]	*Eucommia ulmoides* Oliver wood	Slow pyrolysis	Antibacterial and Antioxidant
Gama et al. [57]	*Eucalyptus urophylla* x *Eucalyptus grandis*, clone I144 wood and *Bambusa vulgaris* culms	Slow pyrolysis	Antibacterial
da Silva et al. [38]	*Eucalyptus urophylla* x *Eucalyptus grandis*, clone I144 wood	Slow pyrolysis	Antibacterial and Cytotoxicity
Theapparat et al. [166]	*Elaeis guineensis* bark	Slow pyrolysis	Antibacterial and atividade cicatrizante
Dias et al. [45]	*Eucalyptus* spp. wood particles	Fast pyrolysis	Antioxidant and Antibacterial
Chenari Bouket et al. [30]	*Pistacia vera*; *Prunus dulcis*; *Pinus* spp.; *Punica granatum*; *Juglans regia* L; *Cupressus* spp. biomass	Slow pyrolysis	Antifungal
Suryani et al. [161]	*Tectona grandis* and *Toona sureni* sawdust	Slow pyrolysis	Antibacterial and Antioxidant
Mansur et al. [97]	*Melaleuca leucadendron* branches	Slow pyrolysis	Antibacterial

(continued)

Table 2 (continued)

Author	Biomass	Type of conversion	Application
Ofoe et al. [116]	*Pinus strobus* biomass	Fast pyrolysis	Antioxidant
Desvita et al. [44]	*Theobroma cacao* L. bark	Slow pyrolysis	Antifungal
Xue et al. [187]	*Eucommia ulmoides* Oliver branches	Slow pyrolysis	Antibacterial, Antifungal and Antioxidant
Sanson et al. [142]	*Eucalyptus* spp. sawdust	Fast pyrolysis	Antifungal
Xin et al. [186]	*Kunzea ericoides* and *Carya tomentosa* wood chips	Fast pyrolysis	Antioxidant
Cheng et al. [31]	*Morus* spp. branches	Slow pyrolysis	Antibacterial and Antifungal
Xin et al. [186]	*Kunzea ericoides* wood chips	Fast pyrolysis	Antibacterial
Li et al. [84]	*Salix psammophila* branches	Slow pyrolysis	Antibacterial
Medeiros and Gasparotto [102]	*Mimosa tenuiflora* and *Eucalyptus urograndis* wood	Slow pyrolysis	Antibacterial
Hu et al. [68]	*Toxicodendron vernicifluum* wood chips	Slow pyrolysis	Antibacterial and Antifungal
Riekkinen et al. [133]	*Pinus sylvestris* sawdust	Slow pyrolysis	Antibacterial and Antifungal
de Souza et al. [40]	*Eucalyptus grandis* wood	Slow pyrolysis	Antioxidant and Cytotoxicity
Takeda et al. [163]	*Acer* spp.; *Quercus* spp.; *Carya* spp.; *Fagus* spp.; *Prunus* spp.	–	Antibacterial
Soares et al. [151]	*Mimosa tenuiflora* and *Eucalyptus Urograndis* wood	Slow pyrolysis	Antioxidant and Cytotoxicity
Baharom et al. [14]	*Averrhoa carambola* and *Mangifera indica* wood	Slow pyrolysis	Antibacterial
Chen et al. [29]	*Eucommia ulmoides* branches	Slow pyrolysis	Antifungal
Li et al. [88]	*Forest residues (Quercus* spp.; *Juglans regia* L.)	Slow pyrolysis	Antioxidant
Suresh et al. [160]	Mixed particles: *Pinus* spp. (80%); *Picea* spp. and *Abies* spp. (20%)	Slow pyrolysis	Antibacterial and Antifungal
Hou et al. [67]H	*Eucommia ulmoides* Olivers wood	Slow pyrolysis	Antibacterial and Antifungal

(continued)

Table 2 (continued)

Author	Biomass	Type of conversion	Application
de Souza Araújo et al. [39]	*Eucalyptus urograndis* and *Mimosa tenuiflora* wood	Slow pyrolysis	Antibacterial and Antifungal
Li et al. [89]	*Quercus* spp. wood	Slow pyrolysis	Antioxidant and Antibacterial
Li et al. [87]	*Larix kaempferi* wood	Slow pyrolysis	Antiviral
Lourençon et al. [92]	*Eucalyptus* spp. wood particles	Fast pyrolysis	Antifungal
Hassan et al. [64]	*Quercus* spp. *and Pinus* spp. wood and bark	Slow pyrolysis	Antioxidant
Patra et al. [123]	*Pinus densiflora* sawdust	Fast pyrolysis	Antibacterial
Theapparat et al. [166]	*Eucalyptus camaldulensis, Leucaena leucocephala, Azadirachta indica, Hevea brasiliensis wood* and *Dendrocalamus asper* culms	Slow pyrolysis	Antifungal
Mmojieje and Hornung [105]	Wood residues	Slow pyrolysis	Antifungal and Antibacterial
Kim et al. [80]	*Liriodendron tulipifera* wood	Fast pyrolysis	Antifungal
Lee et al. [83]	*Quercus acutíssima* wood	Slow pyrolysis	Antifungal
Wei et al. [184]	*Juglans regia* branches	Slow pyrolysis	Antibacterial
Park et al. [122]	*Quercus* spp. wood	–	Antiviral, Antibacterial and Anti-inflammatory
Baimark and Niamsa [15]	*Eucalyptus* spp. wood	Slow pyrolysis	Antifungal
Velmurugan et al. [177]	*Phyllostachys bambusoides*; *Phyllostachys pubescens*; *Phyllostachys nigro*; *Quercus acutíssima and Quercus serrata* wood	Slow pyrolysis	Antifungal
Mohan et al. [107]	*Quercus* spp. and *Pinus* spp. wood and bark	Fast pyrolysis	Antifungal

(continued)

Table 2 (continued)

Author	Biomass	Type of conversion	Application
Ikegami et al. [71]	*Quercus* spp. wood	Slow pyrolysis	Antifungal
Sofos et al. [152]	20 wood species from temperate climate: *Sequoia* spp.; *Carya* spp.; *Acer saccharum*; *Quercus alba*; *Quercus rubra*; *Pinus ponderosa*; *Betula* spp.; *Prosopis* spp.; *Pinus taeda*; *Juniperus* spp.; *Pinus contorta*; *Castanea* spp.; *Prunus* spp.; *Pseudotsuga menziesii*; *Alnus rubra*; *Malus domestica; Juglans nigra*	Slow pyrolysis	Antibacterial

Although the pyrolysis mechanism of forest biomass has been extensively investigated, there are still many challenges and opportunities for the forestry sector. The chemical variability of the biomass and the pyrolysis process variables (temperature, heating rate, and residence time) directly affect the quality, yield, and stability of the pyroligneous liquid. Process optimizations need to be carried out according to the available forest biomass and the required properties of the liquid to meet industrial applications. However, within the concept of biorefinery, the development of technologies and the utilization of pyroligneous liquid for the generation of high-value products, in an economical and sustainable manner, is crucial to strengthening the forestry sector.

6 Agricultural Sector and Potential of Agricultural Biomass

The agricultural sector is crucial for supplying food for human consumption and raw materials widely used across various industries [52, 115]. In addition to meeting these demands, it serves as a significant source of biomass, essential for sectors like bioenergy, biofuels, and bioactives, fostering the development of sustainable and innovative solutions [115]. According to the Food and Agriculture Organization of the United Nations (FAO), between 2000 and 2021, global production of primary crops increased by 54%, reaching 9.5 billion tons—an increase of 2% compared to 2020 [51]. The primary crops produced in 2021 included cereals, sugar crops (22%), vegetables, and oilseeds (12% each). Fruits, roots, and tubers accounted for 9–10% of total production. This growth in agricultural production is driven by the global population's increasing demand. Although many crops can be consumed without processing, certain groups, particularly oilseeds and sugar crops, require processing to produce commodities used as food and fuel. These commodities significantly impact nutrition, health, energy, and the environment [51, 52].

Biomass of biological origin is composed of various components, such as cellulose, hemicellulose, and lignin, as well as extractives, fats, proteins, sugars, moisture, and ash [28, 168]. Although it holds significant potential for use in biorefineries to produce chemicals, biofuels, and other valuable materials, its utilization is still not widely explored [127, 181]. Plant biomass consists of approximately 35–50% cellulose, 20–35% hemicellulose, 15–20% lignin, and 15–20% fats, ash (inorganic minerals such as Ca, K, Fe, Mg, Mn, Al, Si, among others), extractives (waxes, fats, amino acids, pectins), and proteins [112, 168]. Agricultural biomass offers environmental benefits, being renewable, sustainable, and cost-effective. Previous studies have highlighted the significant potential of these residues as raw materials for bio-refineries, enabling the production of chemicals, biofuels, and high-value-added materials [1, 53, 119, 132].

Understanding the chemical composition of agricultural biomass and thermochemical conversion techniques is essential for optimizing pyroligneous liquid production, especially for pharmaceutical applications [93]. The diversity in biomass composition plays a crucial role in the dynamics of conversion processes. Knowledge of these characteristics helps predict how different types of residues react under specific conditions of temperature, heating rate, and residence time, resulting in unique chemical profiles in pyroligneous liquid [165]. Optimizing conversion parameters, such as temperature and residence time, is critical for tailoring pyroligneous liquid profiles and maximizing the production of valuable compounds. This is particularly important in the pharmaceutical industry, where the efficacy and safety of active ingredients are paramount. Directing the production of pyroligneous liquids with specific properties, such as antimicrobial or anti-inflammatory activities, offers a significant competitive advantage.

The composition of biomass varies depending on the plant species, soil conditions, climate, and cultivation practices [46]. To optimize the pyrolysis process and improve the efficiency of thermochemical plants, it is essential to understand the chemical and physicochemical properties of major biomass components like cellulose, hemicellulose, and lignin [132]. Key factors influencing pyroligneous liquid production include: (i) Lignin Content: High elemental carbon content, ranging from 60 to 64%, (ii) Total Biomass Carbon Content: This determines the liquid's energy density and stability [129]. Advancing the understanding of individual biomass components can lead to optimized production of high-value-added bioproducts.

The advancement of industrial processes and the development of new products require solutions that prioritize environmental preservation, reducing reliance on non-renewable resources and promoting the use of biodegradable and sustainable alternatives. In this context, a significant portion of scientific research has focused on developing materials from agricultural and plant-based sources, aiming to explore the potential of these raw materials. However, to maximize process efficiency and the quality of final products, it is essential to understand the chemical characteristics of these biomasses. This knowledge enables the optimization of the conversion processes for each type of raw material, ensuring improved performance during processing. Table 3 highlights the main chemical contributions of different agricultural biomasses reported in the literature.

Table 3 Chemical composition of agricultural biomasses reported in scientific studies

Agricultural biomasses	Components (%)			
	Cellulose	Hemicellulose	Lignin	References
Saccharum officinarum bagasse	45.28	22.13	22.39	Yao et al. [189]
Zea mays bark	75.23	8.99	15.78	Yupa et al. [193]
Oryza sativa straw	39.04	20.91	5.71	Saleem [141]
Manihot esculenta bark	58.40	27.30	6.68	Umriani Permatasari et al. [175]
Sorghum bicolor L. straw	32.0	24.0	13.0	Saini et al. [140]
Hordeum vulgare straw	40.0	30.0	15.0	Saini et al. [140]
Cocos nucifera bark	24.70	12.26	40.10	Cabral et al. [21]
Triticum straw	37–41	27–32	13–15	Saini et al. [140]
Coffea bark	12.50	24.30	9.40	Zeleke et al. [194]
Musa acuminata bark	41.80	38.10	58.50	Supriadi et al. [159]
Glycine max straw	35.00	17.00	21.00	Martelli-Tosi et al. [98]
Arachis hypogaea bark	40.50	14.70	26.40	Saleem [141]
Helianthus annuus bark	30.0	38.50	26.40	De'Nobili et al. [42]
Morus nigra bark	45.11	31.39	17.36	Qiao et al. [126]

6.1 Properties of Pyroligneous Liquid and Its Applications as Bioactives

The properties of pyroligneous liquid are highly relevant, particularly for its applications as a source of bioactive compounds. This liquid, a byproduct of biomass pyrolysis, contains a rich diversity of chemical compounds such as phenols, aldehydes, carboxylic acids, and other phytochemicals. These components exhibit antimicrobial, antioxidant, and anti-inflammatory activities, making pyroligneous liquid a promising resource for industries such as pharmaceuticals and cosmetics [58, 103]. The presence of these compounds makes pyroligneous liquid a promising resource for the pharmaceutical and cosmetic industries, as it can be used in the formulation of medications and natural products aimed at combating infections, reducing inflammation, and promoting overall health. Furthermore, the use of pyroligneous liquid as a bioactive agent not only expands the possibilities for utilizing agricultural residues but also contributes to sustainable practices by transforming low-value materials into products with high therapeutic and functional potential. This duality of waste valorization and the development of innovative products underscores the importance of research and the development of methods to optimize the extraction and application of these bioactive compounds. The main medical and pharmaceutical applications described in the literature are detailed in Table 4.

Table 4 Overview of pyroligneous liquid applications in medical, pharmaceutical, and other utilities reported in the literature

Author	Biomass	Type of conversion	Application
Afrah et al. [1]	*Cocos nucifera bark* and *Opuntia ficus-indica* kernel shells	Slow pyrolysis	Antibacterial and Antioxidant
Ouattara et al. [119]	*Theobroma cacao* and *Coffea* sp. residues	Slow pyrolysis	Antioxidant
Rezzani et al. [132]	*Saccharomyces cerevisiae* celular wall residues	Slow pyrolysis	Antioxidant and Antimicrobial
Yin et al. [191]	*Coptis chinensis*	Slow pyrolysis	Antifungals and Antioxidants
Ratnani et al. [130]	*Pontederia crassipes*	Slow pyrolysis	Antifungals and Antimicrobial
Masfria et al. [100]	*Cocos nucifera* bark	Slow pyrolysis	Antibacterial and Antioxidant
Sari et al. [144]	*Cocos nucifera* bark	Fast pyrolysis	Antibacteriano and Antioxidant
Arundina et al. [10]	*Oryza sativa* bark	Slow pyrolysis	Antioxidant and antiviral
Rajani Kumar and E. Hariprasad [128]	*Cocos nucifera* bark	Slow pyrolysis	Antibacterial and Antioxidant
Novi Caroko and Indah Hartati [114]	*Cocos nucifera* bark	Slow and fast pyrolysis	Antioxidants, antiseptic, antibacterials and food preservation
Anggraini et al. [3]	*Cocos nucifera* bark	Fast pyrolysis	Antimicrobial and Antioxidant
Desvita et al. [43]	*Theobroma cacao* pod shells	Slow pyrolysis	Antimicrobial
Santin et al. [143]	Ocimum *basilicum* plants	Fast pyrolysis	Antioxidant
Mansur et al. [97]	*Oryza sativa* bark	Slow pyrolysis	Antibacterial
Arundina et al. [11]	*Oryza sativa* bark	Slow pyrolysis	Antibacterial
Arundina et al. [9]	*Oryza sativa* bark	Slow pyrolysis	Antimicrobial
Gama et al. [57]	*Bambu vulgaris*	Slow pyrolysis	Antimicrobial and Antibacterial
Suryani et al. [161]	*Theobroma cacao* pod shells	Slow pyrolysis	Antioxidant and Antibacterial
Swastawati et al. [162]	*Cocos nucifera* bark	Slow pyrolysis	Antimicrobial
Kara et al. [77]	*Prunus armeniaca* kernels and *Corylus avellana* bark	Slow pyrolysis	Antifungals and Antioxidants

(continued)

Table 4 (continued)

Author	Biomass	Type of conversion	Application
Vieira et al. [178]	*Cocos nucifera endocarp* and *Enterpe oleracea* seeds	Vacuum pyrolysis	Antimicrobial
Harlina et al. [62]	*Cocos nucifera* bark	Slow pyrolysis	Antioxidant
Ma et al. [95]	*Brassica nopus*	Slow pyrolysis	Antifungal
Wu et al. [185]	*Saccharum officinarum*	Slow pyrolysis	Antibacterial
Chen et al. [26]	*Pennisetum purpureum* Schum	Slow pyrolysis	Anaerobic stability

Pyroligneous liquid, also known as wood vinegar, is a byproduct of lignocellulosic biomass pyrolysis, notably rich in phenolic compounds, organic acids, and other bioactives [3, 43]. The complex composition of this liquid makes it a substance of high potential for a wide range of medical and pharmaceutical applications, as well as use in other industries. In the field of medicine, recent studies have highlighted the effectiveness of pyroligneous liquid in wound healing and tissue regeneration [43]. The antioxidant properties of pyroligneous liquid protect cells from free radical damage, enhancing its use as a bioactive ingredient in pharmaceutical formulations targeting chronic inflammatory and degenerative diseases [128]. Additionally, research indicates that pyroligneous liquid exhibits strong antimicrobial activity against antibiotic-resistant pathogenic bacteria, opening new possibilities for the development of antibacterial therapies for both human and veterinary applications [95, 185].

Outside the medical sector, pyroligneous liquid has demonstrated its utility across various industrial fields. In agriculture, it is widely used as a biofertilizer and natural pesticide. Its ability to function as a biological pest controller stem from the presence of volatile compounds that exert toxic effects on insects and soil-borne pathogenic microorganisms, while simultaneously enhancing the physicochemical properties of the soil, promoting healthy plant growth [119]. The use of this byproduct in sustainable agriculture has been extensively researched as an eco-friendly alternative to synthetic pesticides and chemical fertilizers, contributing to the reduction of the agricultural sector's environmental footprint [144]. In the food industry, pyroligneous liquid is valued for its application in smoking processes, imparting a characteristic flavor to products such as meat, fish, and cheeses. Additionally, it acts as a natural preservative by inhibiting microbial growth, thereby extending the shelf life of food products [191]. Its antioxidant properties also help prevent lipid oxidation in fat-rich foods, such as meats, delaying rancidity and ensuring greater sensory stability [130]. Figure 5 presents the main applications of pyroligneous liquid.

However, despite its wide-ranging applications, the pyrolysis process used to obtain pyroligneous liquid still faces technical and economic challenges. Variability in the chemical composition of biomass, as well as the operational parameters of the pyrolysis process (e.g., temperature, heating rate, and residence time), directly affects

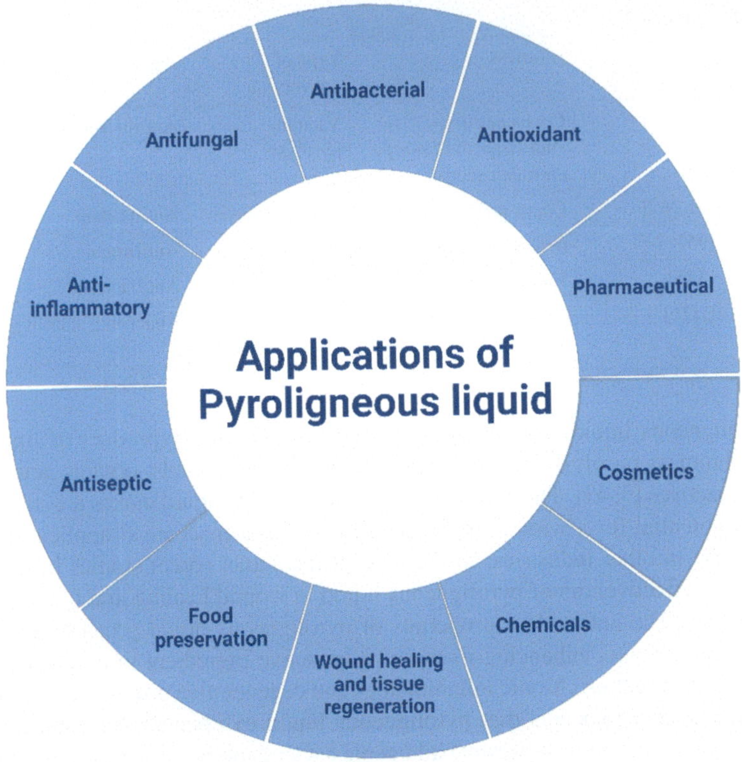

Fig. 5 Main applications of pyroligneous liquid. *Source* Authors (2024)

the quality and yield of the liquid produced [119]. Thus, optimizing the pyrolysis process, particularly for different types of agricultural biomass, is essential to ensure the consistent production of pyroligneous liquids with properties suitable for specific applications [57, 77, 161]. Looking forward, the integration of pyroligneous liquid into biorefinery systems emerges as a promising strategy to enhance the efficiency and sustainability of the production chain. Efforts to develop technologies that maximize the utilization of this byproduct, focusing on generating high-value-added bioactive compounds, will enable the expansion of the agricultural sector and strengthen its competitiveness in the global bioproducts market. This approach promotes a more sustainable and economical use of agricultural biomass [3, 43]. These perspectives highlight that pyroligneous liquid is not only an important industrial innovation but also a cornerstone for the development of new pharmaceutical, medical, and agricultural technologies. Aligned with the concept of a circular bioeconomy, it fosters sustainable development across various sectors.

7 Final Considerations

The world has been facing numerous public health challenges, particularly in underdeveloped and impoverished countries, which have been exacerbated by climate change and adverse environmental factors. The resistance of microorganisms to commonly used drugs for disease treatment has become increasingly concerning, underscoring the importance of searching for new bioactive compounds with antibacterial, antitumor, antioxidant, anti-inflammatory, and other properties. In this context, this chapter clearly presented studies on pyroligneous liquid and its bioactive applications. It is evident that different biomass sources, both forestry and agricultural, can serve as raw materials for producing pyroligneous liquid with bioactive properties. However, the variables of the pyrolysis process and the chemical composition of the biomass can influence the effectiveness of the final product. Despite advancements, it is evident that much research still needs to be conducted in this area. The literature on the cytotoxic, antibiofilm, and anti-inflammatory potential of pyroligneous liquid is nearly nonexistent, and the progression of studies toward possible in vivo applications has not been adequately explored, with most research limited to in vitro testing. Therefore, further studies are essential to investigate these properties and advance knowledge about pyroligneous liquid, thereby contributing to the development of effective and sustainable therapeutic alternatives in the context of increasing drug resistance.

References

1. Afrah BD, Riady MI, Payomthip P, Ramadhanty RV, Rizki F, Alfayyadh ML (2024) Analysis of liquid smoke grade characteristics from coconut shells and palm kernel shell waste through a slow pyrolysis process. J Eng Technol Sci 56(04):545–558. https://doi.org/10.5614/j.eng.technol.sci.2024.56.4.10
2. Aldalin HK, Alharbi NK, Hadi AM, Sharaf M, Mekky AE, Ragab SM, Mahmoud N, Al-Hoshani N, Alwutayd KM, Abdelnour SA (2024) Bioactivity screening and molecular identification of Anchusa milleri L. sunflower crud extract for antioxidant, antiviral, antimicrobial, and anticancer properties. Nat Prod Res. https://doi.org/10.1080/14786419.2024.2390608
3. Anggraini SPA, Suprapto S, Juliastuti SR, Mahfud M (2024) Optimization of pyrolytic oil production from coconut shells by microwave-assisted pyrolysis using activated carbon as a microwave absorber. Int J Renew Energy Dev 13(1):145–157. https://doi.org/10.14710/ijred.2024.56287
4. Anmol, Sharma U (2024) UHPLC-PDA-ELSD-MS-based simultaneous determination of diterpenoid alkaloids in commercially important Himalayan medicinal plant Aconitum heterophyllum Wall. and network pharmacology-based target prediction. Microchem J 205:111210. https://doi.org/10.1016/J.MICROC.2024.111210
5. Antero RVP, Alves ACF, de Oliveira SB, Ojala SA, Brum SS (2020) Challenges and alternatives for the adequacy of hydrothermal carbonization of lignocellulosic biomass in cleaner production systems: a review. J Clean Prod 252:119899. https://doi.org/10.1016/j.jclepro.2019.119899

6. Apiolaza LA, Sharma M (2024) Selection history affects very early expression of wood properties in Pinus radiata. New For (Dordr) 55(4):683–697. https://doi.org/10.1007/s11056-023-09997-3
7. Aria M, Cuccurullo C (2017) Bibliometrix: an R-tool for comprehensive science mapping analysis. J Informetr 11(4):959–975. https://doi.org/10.1016/j.joi.2017.08.007
8. Armanu E-G, Secula MS, Tofanica B-M, Volf I (2024) The impact of biomass composition variability on the char features and yields resulted through thermochemical processes. Polymers (Basel) 16(16):2334. https://doi.org/10.3390/polym16162334
9. Arundina I, Diyatri I, Juliastuti WS, Budhy TI, Surboyo MDC, Iskandar B, Halimah AN, Moelyanto ASA, Ramaniasari SM, Saputra G (2023) Nanoparticles of liquid smoke rice husk inhibit Porphyromonas gingivalis. Eur J Dent 17(02):337–341. https://doi.org/10.1055/s-0042-1749154
10. Arundina I, Frimayanti N, Surboyo MDC, Budhy TI, Iskandar B (2024) 6-Octadecenoic and oleic acid in liquid smoke rice husk showed COVID-19 inhibitor properties. Adv Pharmacol Pharm Sci 2024:1–9. https://doi.org/10.1155/2024/8105595
11. Arundina I, Frimayanti N, Surboyo MDC, Budhy TI, Iskandar B, Pradana A, Rahmaputry T (2023) In silico study of liquid smoke rice husk against COVID-19. Eur J Dent 17(02):492–496. https://doi.org/10.1055/s-0042-1750776
12. Auxenfans T, Buchoux S, Larcher D, Husson G, Husson E, Sarazin C (2014) Enzymatic saccharification and structural properties of industrial wood sawdust: recycled ionic liquids pretreatments. Energy Convers Manage 88:1094–1103. https://doi.org/10.1016/j.enconman.2014.04.027
13. Awasthi MK, Sar T, Gowd SC, Rajendran K, Kumar V, Sarsaiya S, Li Y, Sindhu R, Binod P, Zhang Z, Pandey A, Taherzadeh MJ (2023) A comprehensive review on thermochemical, and biochemical conversion methods of lignocellulosic biomass into valuable end product. Fuel 342:127790. https://doi.org/10.1016/j.fuel.2023.127790
14. Baharom NA, Rahman MHA, Shahrun MS, Suherman FHS, Masdar SNH (2020) Chemical composition and antimicrobial activities of wood vinegars from carambola, coconut shells and mango against selected plant pathogenic microorganisms. Malays J Microbiol https://doi.org/10.21161/mjm.190652
15. Baimark Y, Niamsa N (2009) Study on wood vinegars for use as coagulating and antifungal agents on the production of natural rubber sheets. Biomass Bioenergy 33(6–7):994–998. https://doi.org/10.1016/j.biombioe.2009.04.001
16. Balat M, Balat M, Kırtay E, Balat H (2009) Main routes for the thermo-conversion of biomass into fuels and chemicals. Part 1: pyrolysis systems. Energy Convers Manage 50(12):3147–3157. https://doi.org/10.1016/j.enconman.2009.08.014
17. Bolan S, Padhye LP, Jasemizad T, Govarthanan M, Karmegam N, Wijesekara H, Amarasiri D, Hou D, Zhou P, Biswal BK, Balasubramanian R, Wang H, Siddique KHM, Rinklebe J, Kirkham MB, Bolan N (2024) Impacts of climate change on the fate of contaminants through extreme weather events. Sci Total Environ 909:168388. https://doi.org/10.1016/J.SCITOTENV.2023.168388
18. Boundzanga HM, Cagnon B, Roulet M, de Persis S, Vautrin-Ul C, Bonnamy S (2022) Contributions of hemicellulose, cellulose, and lignin to the mass and the porous characteristics of activated carbons produced from biomass residues by phosphoric acid activation. Biomass Convers Biorefin 12(8):3081–3096. https://doi.org/10.1007/s13399-020-00816-9
19. Bridgwater AV (2012) Review of fast pyrolysis of biomass and product upgrading. Biomass Bioenergy 38:68–94. https://doi.org/10.1016/j.biombioe.2011.01.048
20. Burhenne L, Messmer J, Aicher T, Laborie M-P (2013) The effect of the biomass components lignin, cellulose and hemicellulose on TGA and fixed bed pyrolysis. J Anal Appl Pyrol 101:177–184. https://doi.org/10.1016/j.jaap.2013.01.012
21. Cabral MMS, de Abud AKS, de Silva CEF, Almeida RMRG (2016) Bioethanol production from coconut husk fiber. Ciência Rural 46(10):1872–1877. https://doi.org/10.1590/0103-8478cr20151331

22. Carrillo I, Mendonça RT, Ago M, Rojas OJ (2018) Comparative study of cellulosic components isolated from different Eucalyptus species. Cellulose 25(2):1011–1029. https://doi.org/10.1007/s10570-018-1653-2
23. Cavali M, Soccol CR, Tavares D, Zevallos Torres LA, Oliveira de Andrade Tanobe V, Zandoná Filho A, Woiciechowski AL (2020) Effect of sequential acid-alkaline treatment on physical and chemical characteristics of lignin and cellulose from pine (Pinus spp.) residual sawdust. Bioresour Technol 316:123884. https://doi.org/10.1016/j.biortech.2020.123884
24. Chakhtouna H, Benzeid H, Zari N, Qaiss AEK, Bouhfid R (2024) Recent advances in eco-friendly composites derived from lignocellulosic biomass for wastewater treatment. Biomass Convers Biorefin 14(11):12085–12111. https://doi.org/10.1007/s13399-022-03159-9
25. Chang Z, Ji Y, Sun X, Geng F, Yuan S, Yao X, Ma N, Wu J, Wang K, Song G, Jiang J (2024) Engineering unlocking the synergistic potential of antibacterial wood vinegar and porous activated carbon in the cleaner refining process. Ind Crops Prod 218:118906. https://doi.org/10.1016/j.indcrop.2024.118906
26. Chen D, Zheng M, Zhou Y, Gao L, Zhou W, Wang M, Zhu Y, Xu W (2022) Improving the quality of Napier grass silage with pyroligneous acid: fermentation, aerobic stability, and microbial communities. Front Microbiol 13. https://doi.org/10.3389/fmicb.2022.1034198
27. Chen S, Lin S, Huang S, Cai B, Liang J, Chen Z, Evrendilek F, He Y, Zhong S, Yang Z, Yang C, Liu J (2024) Physicochemical, thermodynamic, and kinetic controls on volatile products, bio-oils, and biochars of Lycium barbarum pyrolysis and their multi-objective optimization. Fuel 367:131522. https://doi.org/10.1016/J.FUEL.2024.131522
28. Chen Y, Fang Y, Yang H, Xin S, Zhang X, Wang X, Chen H (2019) Effect of volatiles interaction during pyrolysis of cellulose, hemicellulose, and lignin at different temperatures. Fuel 248:1–7. https://doi.org/10.1016/j.fuel.2019.03.070
29. Chen Y-H, Li Y-F, Wei H, Li X-X, Zheng H-T, Dong X-Y, Xu T-F, Meng J-F (2020) Inhibition efficiency of wood vinegar on grey mould of table grapes. Food Biosci 38:100755. https://doi.org/10.1016/j.fbio.2020.100755
30. Chenari Bouket A, Narmani A, Sharifi K, Naeimi S, Afshar Mogaddam MR, Hamidi AA, Luptakova L, Alenezi FN, Belbahri L (2023) Semi-VOCs of wood vinegar display strong antifungal activities against oomycete species Globisporangium ultimum and Pythium aphanidermatum. Microbiol Res (Pavia) 14(1):371–389. https://doi.org/10.3390/microbiolres14010029
31. Cheng J, Geng Z-C, Zheng J-L, Qiu L, Jiao F (2022) Characterization of the pyroligneous acids generated from the pyrolysis of four types mulberry branches. Ind Crops Prod 183:114949. https://doi.org/10.1016/j.indcrop.2022.114949
32. Cheng J, Hu SC, Kang K, Li XM, Geng ZC, Zhu MQ (2021) The effects of pyrolysis temperature and storage time on the compositions and properties of the pyroligneous acids generated from cotton stalk based on a polygeneration process. Ind Crops Prod 161:113226. https://doi.org/10.1016/J.INDCROP.2020.113226
33. Cheruvari A, Kammara R (2025) Bacteriocins future perspectives: substitutes to antibiotics. Food Control 168:110834. https://doi.org/10.1016/J.FOODCONT.2024.110834
34. Clemente-Castro S, Palma A, Ruiz-Montoya M, Giráldez I, Díaz MJ (2023) Comparative study of the combustion, pyrolysis and gasification processes of Leucaena leucocephala: kinetics and gases obtained. Heliyon 9(7):2405–8440. https://doi.org/10.1016/J.HELIYON.2023.E17943
35. Collard F-X, Blin J (2014) A review on pyrolysis of biomass constituents: mechanisms and composition of the products obtained from the conversion of cellulose, hemicelluloses and lignin. Renew Sustain Energy Rev 38:594–608. https://doi.org/10.1016/j.rser.2014.06.013
36. Cupertino GFM, Pereira AKS, Ucella-Filho JGM, Delatorre FM, da Silva ÁM, da Silva KCA, de Souza EC, Parreira LA, Pimenta AS, Saloni D, Luque R, Dias Júnior AF (2024) Prospects for the utilization of bio-oil-derived chemicals generated via co-pyrolysis of biomass and polyethylene terephthalate (PET). J Anal Appl Pyrol 183:106781. https://doi.org/10.1016/J.JAAP.2024.106781

37. Cupertino GFM, da Silva ÁM, Pereira AKS, Delatorre FM, Ucella-Filho JGM, de Souza EC, Profeti D, Profeti LPR, Oliveira MP, Saloni D, Luque R, Dias Júnior AF (2024) Co-pyrolysis of biomass and polyethylene terephthalate (PET) as an alternative for energy production from waste valorization. Fuel 362:130761. https://doi.org/10.1016/J.FUEL.2023.130761
38. da Silva BA, Feijó FMC, Alves ND, Pimenta AS, Benicio LDM, da Silva Júnior EC, Santos CS, Pereira AF, Moura YBF, Gama GSP, de Neto ESL (2023) Use of a product based on wood vinegar of Eucalyptus clone I144 used in the control of bovine mastitis. Vet Microbiol 279:109670. https://doi.org/10.1016/j.vetmic.2023.109670
39. de Souza AE, Pimenta AS, Feijó FMC, Castro RVO, Fasciotti M, Monteiro TVC, de Lima KMG (2018) Antibacterial and antifungal activities of pyroligneous acid from wood of *Eucalyptus urograndis* and *Mimosa tenuiflora*. J Appl Microbiol 124(1):85–96. https://doi.org/10.1111/jam.13626
40. de Souza JLS, Alves T, Camerini L, Nedel F, Campos AD, Lund RG (2021) Antimicrobial and cytotoxic capacity of pyroligneous extracts films of Eucalyptus grandis and chitosan for oral applications. Sci Rep 11(1):21531. https://doi.org/10.1038/s41598-021-00529-7
41. de Souza JLS, da Guimarães VBS, Campos AD, Lund RG (2018) Antimicrobial potential of pyroligneous extracts–a systematic review and technological prospecting. Braz J Microbiol 49(Suppl 1):128–139. https://doi.org/10.1016/J.BJM.2018.07.001
42. De'Nobili MD, Bernhardt DC, Basanta MF, Rojas AM (2021) Sunflower (Helianthus annuus L.) seed hull waste: composition, antioxidant activity, and filler performance in pectin-based film composites. Front Nutr 8. https://doi.org/10.3389/fnut.2021.777214
43. Desvita H, Faisal M, Mahidin M, Suhendrayatna (2023) Natural antimicrobial properties of liquid smoke derived from cocoa pod shells in meatball preservation. S Afr J Chem Eng 46:106–111. https://doi.org/10.1016/j.sajce.2023.08.003
44. Desvita H, Faisal M, Mahidin S (2022) Antimicrobial potential of wood vinegar from cocoa pod shells (Theobroma cacao L.) against Candida albicans and Aspergillus niger. Mater Today Proc 63:S210–S213. https://doi.org/10.1016/j.matpr.2022.02.410
45. Dias IA, Horta RP, Matos M, Helm CV, Magalhães WLE, de Lima EA, da Silva BJG, de Muniz GIB, de Cademartori PHG (2024) Exploring the antioxidant and antimicrobial properties of the water-soluble fraction derived from pyrolytic lignin separation in fast-pyrolysis bio-oil. Biomass Convers Biorefin 14(19):24333–24344. https://doi.org/10.1007/s13399-023-04561-7
46. Díez D, Urueña A, Piñero R, Barrio A, Tamminen T (2020) Determination of hemicellulose, cellulose, and lignin content in different types of biomasses by thermogravimetric analysis and pseudocomponent kinetic model (TGA-PKM method). Processes 8(9):1048. https://doi.org/10.3390/pr8091048
47. Dusdal J, Powell JJW (2021) Benefits, motivations, and challenges of international collaborative research: a sociology of science case study. Sci Public Policy 48(2):235–245. https://doi.org/10.1093/scipol/scab010
48. Efika CE, Onwudili JA, Williams PT (2018) Influence of heating rates on the products of high-temperature pyrolysis of waste wood pellets and biomass model compounds. Waste Manage 76:497–506. https://doi.org/10.1016/J.WASMAN.2018.03.021
49. El Bari H, Fanezoune CK, Dorneanu B, Arellano-Garcia H, Majozi T, Elhenawy Y, Bayssi O, Hirt A, Peixinho J, Dhahak A, Gadalla MA, Khashaba NH, Ashour FH (2024) Catalytic fast pyrolysis of lignocellulosic biomass: recent advances and comprehensive overview. J Anal Appl Pyrol 178:106390. https://doi.org/10.1016/j.jaap.2024.106390
50. El-Fawy MM, Abo-Elyousr KAM, Sallam NMA, El-Sharkawy RMI, Ibrahim YE (2023) Fungicidal effect of guava wood vinegar against Colletotrichum coccodes causing black dot disease of potatoes. Horticulturae 9(6):710. https://doi.org/10.3390/horticulturae9060710
51. FAO F and AO of the UN (2023) World food and agriculture—statistical pocketbook 2023. FAO, Rome, Italy
52. Faqir Y, Qayoom A, Erasmus E, Schutte-Smith M, Visser HG (2024) A review on the application of advanced soil and plant sensors in the agriculture sector. Comput Electron Agric 226:109385. https://doi.org/10.1016/j.compag.2024.109385

53. Fedeli R, Dichiara M, Carullo G, Tudino V, Gemma S, Butini S, Campiani G, Loppi S (2024) Unlocking the potential of biostimulants in sustainable agriculture: effect of wood distillate on the nutritional profiling of apples. Heliyon 10(18):e37599. https://doi.org/10.1016/j.heliyon.2024.e37599
54. Feng T-H, Xue R, Zhu L, Zhu Y-H (2023) Comparison of antioxidant and antibacterial effects of the Eucommia wood vinegar under the refining methods of atmospheric distillation and vacuum distillation. Ind Crops Prod 192:116013. https://doi.org/10.1016/j.indcrop.2022.116013
55. Fisher MC, Alastruey-Izquierdo A, Berman J, Bicanic T, Bignell EM, Bowyer P, Bromley M, Brüggemann R, Garber G, Cornely OA, SarahJ G, Harrison TS, Kuijper E, Rhodes J, Sheppard DC, Warris A, White PL, Xu J, Zwaan B, Verweij PE (2022) Tackling the emerging threat of antifungal resistance to human health. Nat Rev Microbiol 20(9):557–571. https://doi.org/10.1038/s41579-022-00720-1
56. Franco M, Pinho C (2019) A case study about cooperation between University Research Centres: knowledge transfer perspective. J Innov Knowl 4(1):62–69. https://doi.org/10.1016/j.jik.2018.03.003
57. Gama GSP, Pimenta AS, Feijó FMC, Santos CS, Fernandes BCC, de Oliveira MF, de Souza EC, Monteiro TVC, Fasciotti M, de Azevedo TKB, de Melo RR, Júnior AFD (2023) Antimicrobial activity and chemical profile of wood vinegar from eucalyptus (Eucalyptus urophylla x Eucalyptus grandis—clone I144) and bamboo (Bambusa vulgaris). World J Microbiol Biotechnol 39(7):186. https://doi.org/10.1007/s11274-023-03628-x
58. Grewal A, Abbey L, Gunupuru LR (2018) Production, prospects and potential application of pyroligneous acid in agriculture. J Anal Appl Pyrol 135:152–159. https://doi.org/10.1016/j.jaap.2018.09.008
59. Grzybek P, Dudek G, van der Bruggen B (2024) Cellulose-based films and membranes: a comprehensive review on preparation and applications. Chem Eng J 495:153500. https://doi.org/10.1016/j.cej.2024.153500
60. Guedes RE, Luna AS, Torres AR (2018) Operating parameters for bio-oil production in biomass pyrolysis: a review. J Anal Appl Pyrol 129:134–149. https://doi.org/10.1016/J.JAAP.2017.11.019
61. Güleç F, Parthiban A, Umenweke GC, Musa U, Williams O, Mortezaei Y, Suk-Oh H, Lester E, Ogbaga CC, Gunes B, Okolie JA (2024) Progress in lignocellulosic biomass valorization for biofuels and value-added chemical production in the <scp>EU</scp>: a focus on thermochemical conversion processes. Biofuels Bioprod Biorefin 18(3):755–781. https://doi.org/10.1002/bbb.2544
62. Harlina PW, Yuliana T, Fetriyuna SR, Ma M (2023) Study on the development and functional characteristics of salted egg with liquid smoke. Food Sci Anim Resour 43(3):471–490. https://doi.org/10.5851/kosfa.2023.e10
63. Harris N, Gibbs D (2021) Forests absorb twice as much carbon as they emit each year. https://www.wri.org/insights/forests-absorb-twice-much-carbon-they-emit-each-year. Accessed 6 Oct 2024
64. Hassan EB, El-Giar EM, Steele P (2016) Evaluation of the antioxidant activities of different bio-oils and their phenolic distilled fractions for wood preservation. Int Biodeterior Biodegradation 110:121–128. https://doi.org/10.1016/j.ibiod.2016.03.015
65. Heo S, Chan AY, Diaz Peralta P, Jin L, Pereira Nunes CR, Bell ML (2022) Impacts of the COVID-19 pandemic on scientists' productivity in science, technology, engineering, mathematics (STEM), and medicine fields. Humanit Soc Sci Commun 9(1):434. https://doi.org/10.1057/s41599-022-01466-0
66. Herrera K, Morales LF, Tarazona NA, Aguado R, Saldarriaga JF (2022) Use of biochar from rice husk pyrolysis: part A: recovery as an adsorbent in the removal of emerging compounds. ACS Omega 7(9):7625–7637. https://doi.org/10.1021/ACSOMEGA.1C06147/ASSET/IMAGES/MEDIUM/AO1C06147_M005.GIF
67. Hou X, Qiu L, Luo S, Kang K, Zhu M, Yao Y (2018) Chemical constituents and antimicrobial activity of wood vinegars at different pyrolysis temperature ranges obtained from *Eucommia*

ulmoides Olivers branches. RSC Adv 8(71):40941–40949. https://doi.org/10.1039/C8RA07491G
68. Hu S-C, Cheng J, Wang W-P, Zhu Y-H, Kang K, Zhu M-Q, Huang X-H (2022) Preparation and analysis of pyroligneous liquor, charcoal and gas from lacquer wood by carbonization method based on a biorefinery process. Energy 239:121918. https://doi.org/10.1016/j.energy.2021.121918
69. Hu X, Wang Y, Mourant D, Gunawan R, Lievens C, Chaiwat W, Gholizadeh M, Wu L, Li X, Li CZ (2013) Polymerization on heating up of bio-oil: a model compound study. AIChE J 59(3):888–900. https://doi.org/10.1002/AIC.13857
70. Ighalo JO, Iwuchukwu FU, Eyankware OE, Iwuozor KO, Olotu K, Bright OC, Igwegbe CA (2022) Flash pyrolysis of biomass: a review of recent advances. Clean Technol Environ Policy 24(8):2349–2363. https://doi.org/10.1007/s10098-022-02339-5
71. Ikegami F, Sekine T, Fujii Y (1998) Anti-dermaptophyte activity of phenolic compounds in "Mokusaku-Eki." Yakugaku Zasshi 118(1):27–30. https://doi.org/10.1248/yakushi1947.118.1_27
72. IBÁ, Indústria Brasileira de Árvores (2024) Relatório anual
73. Jeeru LR, Abdul FK, Anireddy JS, Ch VP, Dhanavath KN (2023) Optimization of process parameters for conventional pyrolysis of algal biomass into bio–oil and bio–char production. Chem Eng Process Process Intensification 185:109311. https://doi.org/10.1016/J.CEP.2023.109311
74. Jeffri NI, Mohammad Rawi NF, Mohamad Kassim MH, Abdullah CK (2024) Unlocking the potential: evolving role of technical lignin in diverse applications and overcoming challenges. Int J Biol Macromol 274:133506. https://doi.org/10.1016/j.ijbiomac.2024.133506
75. Jo J-H, Choi M, Kweon D, Son Y-G, Lim EM (2024) Regulating ecosystem services in a local forest: navigating supply, trade-offs, and synergies. Trees Forest People 15:100466. https://doi.org/10.1016/j.tfp.2023.100466
76. Justicia J, Alberto Baeza J, de Oliveira AS, Calvo L, Heras F, Gilarranz MA (2022) Aqueous-phase reforming of water-soluble compounds from pyrolysis bio-oils. Renew Energy 199:895–907. https://doi.org/10.1016/J.RENENE.2022.09.021
77. Kara M, Soylu S, Soylu EM, Uysal A, Kurt Ş, Türkmen M (2024) Determination of the chemical composition and antifungal activity of wood vinegar (Pyroligneous acid) against the onion bulb rot disease caused by Fusarium proliferatum. J Crop Health 76(1):75–85. https://doi.org/10.1007/s10343-023-00931-3
78. Kartal SN, Imamura Y, Tsuchiya F, Ohsato K (2004) Preliminary evaluation of fungicidal and termiticidal activities of filtrates from biomass slurry fuel production. Bioresour Technol 95(1):41–47. https://doi.org/10.1016/j.biortech.2004.02.005
79. KC YB, Liu Q, Saud P, Gaire D, Adhikari H (2024) Estimation of Above-Ground Forest Biomass in Nepal by the Use of Airborne LiDAR, and Forest Inventory Data. Land (Basel) 13(2):213. https://doi.org/10.3390/land13020213
80. Kim KH, Jeong HS, Kim J-Y, Han GS, Choi I-G, Choi JW (2012) Evaluation of the antifungal effects of bio-oil prepared with lignocellulosic biomass using fast pyrolysis technology. Chemosphere 89(6):688–693. https://doi.org/10.1016/j.chemosphere.2012.06.010
81. Kovats RS, Campbell-Lendrum DH, Mcmichael AJ, Woodward A, St J, Cox H (2001) Early effects of climate change: do they include changes in vector-borne disease? Philos Trans Roy Soc B Biol Sci 356(1411):1057. https://doi.org/10.1098/RSTB2001.0894
82. Le Floch A, Jourdes M, Teissedre P-L (2015) Polysaccharides and lignin from oak wood used in cooperage: composition, interest, assays: a review. Carbohydr Res 417:94–102. https://doi.org/10.1016/j.carres.2015.07.003
83. Lee CS, Yi EH, Kim H-R, Huh S-R, Sung S-H, Chung M-H, Ye S-K (2011) Anti-dermatitis effects of oak wood vinegar on the DNCB-induced contact hypersensitivity via STAT3 suppression. J Ethnopharmacol 135(3):747–753. https://doi.org/10.1016/j.jep.2011.04.009
84. Li J, Ma X, Duan H (2022) Preparation, chemical constituents and antimicrobial activity of pyroligneous acids from Salix psammophila branches. Wood Res 67(1):1–10. https://doi.org/10.37763/wr.1336-4561/67.1.0110

85. Li J, Xu K, Yao X, Liu J (2024) Investigation of biomass slow pyrolysis mechanisms based on the generation trends in pyrolysis products. Process Saf Environ Prot 183:327–338. https://doi.org/10.1016/j.psep.2024.01.027
86. Li L, Rowbotham JS, Christopher Greenwell H, Dyer PW (2013) An introduction to pyrolysis and catalytic pyrolysis: versatile techniques for biomass conversion. In: New and future developments in catalysis. Elsevier, pp 173–208
87. Li R, Narita R, Ouda R, Kimura C, Nishimura H, Yatagai M, Fujita T, Watanabe T (2018) Structure-dependent antiviral activity of catechol derivatives in pyroligneous acid against the encephalomycarditis virus. RSC Adv 8(63):35888–35896. https://doi.org/10.1039/C8RA07096B
88. Li Z, Wu L, Sun S, Gao J, Zhang H, Zhang Z, Wang Z (2019) Disinfection and removal performance for Escherichia coli, toxic heavy metals and arsenic by wood vinegar-modified zeolite. Ecotoxicol Environ Saf 174:129–136. https://doi.org/10.1016/j.ecoenv.2019.01.124
89. Li Z, Zhang L, Chen G, Wu L, Liu B, Li Y, Sun S, Zhang H, Zhang Z, Wang Z (2018) A new method for comprehensive utilization of wood vinegar by distillation and liquid−liquid extraction. Process Biochem 75:194–201. https://doi.org/10.1016/j.procbio.2018.08.012
90. Liao G, Sun E, Kana EBG, Huang H, Sanusi IA, Qu P, Jin H, Liu J, Shuai L (2024) Renewable hemicellulose-based materials for value-added applications. Carbohydr Polym 341:122351. https://doi.org/10.1016/j.carbpol.2024.122351
91. Liu Q, Zhou L, Xie X, Fan D, Ouyang X, Fan W, Qiu X (2022) Enhanced production and separation of short-chain glucan oligomers from corn stover in an unacidified LiBr molten salt hydrate *via* pre-extraction of hemicellulose. Green Chem 24(22):8812–8819. https://doi.org/10.1039/D2GC03396H
92. Lourençon TV, Mattos BD, Cademartori PHG, Magalhães WLE (2016) Bio-oil from a fast pyrolysis pilot plant as antifungal and hydrophobic agent for wood preservation. J Anal Appl Pyrol 122:1–6. https://doi.org/10.1016/j.jaap.2016.11.004
93. Ma C, Li W, Zu Y, Yang L, Li J (2014) Antioxidant properties of pyroligneous acid obtained by thermochemical conversion of Schisandra chinensis Baill. Molecules 19(12):20821–20838. https://doi.org/10.3390/MOLECULES191220821
94. Ma C, Song K, Yu J, Yang L, Zhao C, Wang W, Zu G, Zu Y (2013) Pyrolysis process and antioxidant activity of pyroligneous acid from Rosmarinus officinalis leaves. J Anal Appl Pyrol 104:38–47. https://doi.org/10.1016/j.jaap.2013.09.011
95. Ma J, Islam F, Ayyaz A, Fang R, Hannan F, Farooq MA, Ali B, Huang Q, Sun R, Zhou W (2022) Wood vinegar induces salinity tolerance by alleviating oxidative damages and protecting photosystem II in rapeseed cultivars. Ind Crops Prod 189:115763. https://doi.org/10.1016/j.indcrop.2022.115763
96. Malico I, Gonçalves AC (2024) Sources and distribution of forest biomass for energy, pp 25–52
97. Mansur D, Sugiwati S, Rizal WA, Suryani R, Maryana R (2023) Pyrolysis of cajuput (Melaleuca leucadendron) twigs and rice (Oryza sativa) husks to produce liquid smoke-containing fine chemicals for antibacterial agent application. Biomass Convers Biorefin 13(12):10561–10574. https://doi.org/10.1007/s13399-021-01896-x
98. Martelli-Tosi M, da Torricillas MS, Martins MA, de Assis OBG, Tapia-Blácido DR (2016) Using commercial enzymes to produce cellulose nanofibers from soybean straw. J Nanomater 2016:1–10. https://doi.org/10.1155/2016/8106814
99. Martins-Vieira D, Missau JC, Anshu J, Siakpebru K, Thengane OK, Morais SK, Tanabe ARC, Bertuol EH, Review DA, Lachos-Perez D, Cláudio Martins-Vieira J, Missau J, Anshu K, Siakpebru OK, Thengane SK, Rita A, Morais C, Tanabe EH, Assumpção Bertuol D (2023) Review on biomass pyrolysis with a focus on bio-oil upgrading techniques. Analytica 4(2):182–205. https://doi.org/10.3390/ANALYTICA4020015
100. Masfria, Syahputra H, Wahyuni HS, Zebua NF, Elizabeth, Marcelynn (2024) Analysis of total flavonoid and antioxidant activity of coconut shell liquid smoke (Cocos nucifera L.) as an antibacterial. Pharm Educ 24(2):39–45. https://doi.org/10.46542/pe.2024.242.3945

101. McMichael C, Dasgupta S, Ayeb-Karlsson S, Kelman I (2020) A review of estimating population exposure to sea-level rise and the relevance for migration. Environ Res Lett 15(12):123005. https://doi.org/10.1088/1748-9326/ABB398
102. Medeiros LCD, Gasparotto LHS (2022) Pyroligneous acid and antibacterial activity: criticism of a paper by Araújo et al. (2018). J Appl Microbiol 132(3):1768–1770. https://doi.org/10.1111/jam.15281
103. Miethke M, Pieroni M, Weber T, Brönstrup M, Hammann P, Halby L, Arimondo PB, Glaser P, Aigle B, Bode HB, Moreira R, Li Y, Luzhetskyy A, Medema MH, Pernodet JL, Stadler M, Tormo JR, Genilloud O, Truman AW, Weissman KJ, Takano E, Sabatini S, Stegmann E, Brötz-Oesterhelt H, Wohlleben W, Seemann M, Empting M, Hirsch AKH, Loretz B, Lehr CM, Titz A, Herrmann J, Jaeger T, Alt S, Hesterkamp T, Winterhalter M, Schiefer A, Pfarr K, Hoerauf A, Graz H, Graz M, Lindvall M, Ramurthy S, Karlén A, van Dongen M, Petkovic H, Keller A, Peyrane F, Donadio S, Fraisse L, Piddock LJV, Gilbert IH, Moser HE, Müller R (2021) Towards the sustainable discovery and development of new antibiotics. Nat Rev Chem 5(10):726–749. https://doi.org/10.1038/s41570-021-00313-1
104. Millati R, Cahyono RB, Ariyanto T, Azzahrani IN, Putri RU, Taherzadeh MJ (2019) Agricultural, industrial, municipal, and forest wastes. In: Sustainable resource recovery and zero waste approaches. Elsevier, pp 1–22
105. Mmojieje J, Hornung A (2015) The potential application of pyroligneous acid in the UK agricultural industry. J Crop Improv 29(2):228–246. https://doi.org/10.1080/15427528.2014.995328
106. Mohan D, Pittman CU, Steele PH (2006) Pyrolysis of wood/biomass for bio-oil: a critical review. Energy Fuels 20(3):848–889. https://doi.org/10.1021/EF0502397/ASSET/IMAGES/LARGE/EF0502397H00010.JPEG
107. Mohan D, Shi J, Nicholas DD, Pittman CU, Steele PH, Cooper JE (2008) Fungicidal values of bio-oils and their lignin-rich fractions obtained from wood/bark fast pyrolysis. Chemosphere 71(3):456–465. https://doi.org/10.1016/j.chemosphere.2007.10.049
108. Mohd Hamzah MAA, Hasham R, Nik Malek NAN, Hashim Z, Yahayu M, Abdul Razak FI, Zakaria ZA (2022) Beyond conventional biomass valorisation: pyrolysis-derived products for biomedical applications. Biofuel Res J 9(3):1648–1658. https://doi.org/10.18331/BRJ2022.9.3.2
109. Mourant D, Lievens C, Gunawan R, Wang Y, Hu X, Wu L, Syed-Hassan SSA, Li CZ (2013) Effects of temperature on the yields and properties of bio-oil from the fast pyrolysis of mallee bark. Fuel 108:400–408. https://doi.org/10.1016/J.FUEL.2012.12.018
110. Mujtaba M, Fernandes Fraceto L, Fazeli M, Mukherjee S, Savassa SM, de Medeiros AG, do Espírito Santo Pereira A, Mancini SD, Lipponen J, Vilaplana F (2023) Lignocellulosic biomass from agricultural waste to the circular economy: a review with focus on biofuels, biocomposites and bioplastics. J Clean Prod 402:136815. https://doi.org/10.1016/J.JCLEPRO.2023.136815
111. Naqvi SR, Tariq R, Shahbaz M, Naqvi M, Aslam M, Khan Z, Mackey H, Mckay G, Al-Ansari T (2021) Recent developments on sewage sludge pyrolysis and its kinetics: resources recovery, thermogravimetric platforms, and innovative prospects. Comput Chem Eng 150:107325. https://doi.org/10.1016/J.COMPCHEMENG.2021.107325
112. Nizamuddin S, Baloch HA, Siddiqui MTH, Mubarak NM, Tunio MM, Bhutto AW, Jatoi AS, Griffin GJ, Srinivasan MP (2018) An overview of microwave hydrothermal carbonization and microwave pyrolysis of biomass. Rev Environ Sci Biotechnol 17(4):813–837. https://doi.org/10.1007/s11157-018-9476-z
113. Nolander C, Lundmark R (2024) A review of forest ecosystem services and their spatial value characteristics. Forests 15(6):919. https://doi.org/10.3390/f15060919
114. Novi C, Indah H (2024) Characterization of Distillation and redistillation product of coconut shell liquid smoke at various temperatures. J Adv Res Fluid Mech Therm Sci 113(1):44–56. https://doi.org/10.37934/arfmts.113.1.4456
115. Nsabiyeze A, Ma R, Li J, Luo H, Zhao Q, Tomka J, Zhang M (2024) Tackling climate change in agriculture: a global evaluation of the effectiveness of carbon emission reduction policies. J Clean Prod 468:142973. https://doi.org/10.1016/j.jclepro.2024.142973

116. Ofoe R, Gunupuru LR, Abbey L (2022) Metabolites, elemental profile and chemical activities of Pinus strobus high temperature-derived pyroligneous acid. Chem Biol Technol Agric 9(1):85. https://doi.org/10.1186/s40538-022-00357-5
117. Ojo AB, Adanlawo IG (2024) Antioxidant, antidiabetic, and anti-inflammatory activities of flavonoid-rich fractions of Solanum anguivi Lam. fruit: In vitro and ex vivo studies. Heliyon 10(11):e31895. https://doi.org/10.1016/J.HELIYON.2024.E31895
118. Oramahi HA, Maurisa T, Nurhaida N, Darwati H, Rifanjani S (2024) Optimization and characterization of liquid smoke produced by Terminalia catappa wood pyrolysis and its in vitro antifungal activity. Sci Technol Indonesia 9(1):207–214. https://doi.org/10.26554/sti.2024.9.1.207-214
119. Ouattara H, Niamke BF, Ahoke MEMF, N'guessan JLL, Yao J, Amusant N, Dumarcay S (2024a) Physicochemical characterization, chemical composition and antioxidant activities of pyroligneous acids from cocktails of wood and agricultural residues from Ivory Coast. J Anal Appl Pyrol 181:106629. https://doi.org/10.1016/j.jaap.2024.106629
120. Ouattara HAA, Niamké FB, Yao JC, Amusant N, Garnier B (2023) Wood vinegars: production processes, properties, and valorization. For Prod J 73(3):239–249. https://doi.org/10.13073/FPJ-D-23-00021
121. Panwar NL, Pawar A, Salvi BL (2019) Comprehensive review on production and utilization of biochar. SN Appl Sci 1(2):1–19. https://doi.org/10.1007/S42452-019-0172-6/FIGURES/9
122. Park S-J, Lee A, Back S-K, Youn H-S (2010) Suppression of TRIF-dependent signaling pathway of toll-like receptors by oak wood vinegar in RAW264.7 macrophages. Mol Cell Toxicol 6(1):73–78. https://doi.org/10.1007/s13273-010-0010-9
123. Patra JK, Hwang H, Choi JW, Baek K-H (2015) Bactericidal mechanism of bio-oil obtained from fast pyrolysis of *Pinus densiflora* against two foodborne pathogens, *Bacillus cereus* and *Listeria monocytogenes*. Foodborne Pathog Dis 12(6):529–535. https://doi.org/10.1089/fpd.2014.1914
124. Patz JA, Campbell-Lendrum D, Holloway T, Foley JA (2005) Impact of regional climate change on human health. Nature 438(7066):310–317. https://doi.org/10.1038/nature04188
125. Pereira EG, Fauller H, Magalhães M, Guirardi B, Martins MA (2022) Potential use of wood pyrolysis coproducts: a review. Environ Prog Sustain Energy 41(1). https://doi.org/10.1002/ep.13705
126. Qiao W, Tao J, Luo Y, Tang T, Miao J, Yang Q (2018) Microbial oil production from solid-state fermentation by a newly isolated oleaginous fungus, *Mucor circinelloides* Q531 from mulberry branches. R Soc Open Sci 5(11):180551. https://doi.org/10.1098/rsos.180551
127. Qureshi SS, Nizamuddin S, Xu J, Vancov T, Chen C (2024) Cellulose nanocrystals from agriculture and forestry biomass: synthesis methods, characterization and industrial applications. Environ Sci Pollut Res. https://doi.org/10.1007/s11356-024-35127-3
128. Rajani Kumar K, Hariprasad E (2024) Simultaneous determination of rilpivirine and cabotegravir in bulk and pharmaceutical dosage form: stability indicating RP-HPLC method development and validation. RASAYAN J Chem 17(01):259–266. https://doi.org/10.31788/RJC.2024.1718564
129. Rashid T, Sher F, Khan AS, Khalid U, Rasheed T, Iqbal HMN, Murugesan T (2021) Effect of protic ionic liquid treatment on the pyrolysis products of lignin extracted from oil palm biomass. Fuel 291:120133. https://doi.org/10.1016/j.fuel.2021.120133
130. Ratnani RD, Arianti FD, Sasongko NA (2024) Exploring the potential of water hyacinth weed (Pontederia crassipes) as an environmentally friendly antifungal to realize sustainable development in lakes: a review. Case Stud Chem Environ Eng 9:100702. https://doi.org/10.1016/j.cscee.2024.100702
131. Reddy VP (2023) Oxidative stress in health and disease. Biomedicines 11(11):2925. https://doi.org/10.3390/biomedicines11112925
132. Rezzani GD, Salvay AG, Peltzer MA (2024) Bioactive materials from brewery yeast cell wall residues with the addition of refined liquid smoke: development and study of functional properties. J Polym Environ 32(7):3285–3297. https://doi.org/10.1007/s10924-023-03165-6

133. Riekkinen K, Raninen K, Keränen E, Selenius M, Vilppo T, Raatikainen O, Korhonen J (2022) Antimicrobial activity of slow pyrolysis distillates from pine wood biomass against three pathogens. Forests 13(4):559. https://doi.org/10.3390/f13040559
134. Rizal WA, Wahono SK, Yuliansyah AT (2024) Enhancing wood vinegar quality: integrated pyrolysis process of cajuput (Melaleuca leucadendron) twigs biomass with gas-phase adsorption using zeolite adsorbents. Bioresour Technol Rep 26:101840. https://doi.org/10.1016/j.biteb.2024.101840
135. Rocha KC, Alonso CG, Leal WGO, Schultz EL, Andrade LA, Ostroski IC (2020) Slow pyrolysis of Spirulina platensis for the production of nitrogenous compounds and potential routes for their separation. Bioresour Technol 313:123709. https://doi.org/10.1016/J.BIORTECH.2020.123709
136. Rocklöv J, Dubrow R (2020) Climate change: an enduring challenge for vector-borne disease prevention and control (Nature Immunology, (2020), 21(5):479–483), https://doi.org/10.1038/s41590-020-0648-y). Nat Immunol 21(6):695. https://doi.org/10.1038/S41590-020-0692-7
137. Rodilla SD, Martínez-Pineda M, Yagüe-Ruiz C, Vercet A (2023) Evaluation of phenolic compounds, antioxidant activity and pigment content in emerging and traditional plant-based oils in Mediterranean gastronomy. Int J Gastron Food Sci 33:100771. https://doi.org/10.1016/J.IJGFS.2023.100771
138. Rosado MJ, Rencoret J, Gutiérrez A, del Río JC (2023) Structural characterization of the milled-wood lignin isolated from sweet orange tree (Citrus sinensis) pruning residue. Polymers (Basel) 15(8):1840. https://doi.org/10.3390/polym15081840
139. Saberi Riseh R, Fathi F, Lagzian A, Vatankhah M, Kennedy JF (2024) Modifying lignin: a promising strategy for plant disease control. Int J Biol Macromol 271:132696. https://doi.org/10.1016/j.ijbiomac.2024.132696
140. Saini JK, Saini R, Tewari L (2015) Lignocellulosic agriculture wastes as biomass feedstocks for second-generation bioethanol production: concepts and recent developments. 3 Biotech 5(4):337–353. https://doi.org/10.1007/s13205-014-0246-5
141. Saleem M (2022) Possibility of utilizing agriculture biomass as a renewable and sustainable future energy source. Heliyon 8(2):e08905. https://doi.org/10.1016/j.heliyon.2022.e08905
142. Sanson CSL, Helm CV, Magalhães WLE, de Muniz GIB, Missio AL, de Cademartori PHG (2022) Enhancing water repellency and decay resistance of wood by using water-soluble fractions separated from pyrolytic lignin of fast-pyrolysis bio-oil. Ind Crops Prod 177:114540. https://doi.org/10.1016/j.indcrop.2022.114540
143. Santin M, Becagli M, Sciampagna MC, Mannucci A, Ranieri A, Castagna A (2023) Integrative effect of UV-B and some organic amendments on growth, phenolic and flavonoid compounds, and antioxidant activity of basil (Ocimum basilicum L.) plants. Horticulturae 9(8):894. https://doi.org/10.3390/horticulturae9080894
144. Sari RM, Wahono SK, Anwar M, Anggo Rizal W, Suryani R, Suwanto A (2024) Wet scrubbing pre-condensation purification of wood vinegar from pyrolysis coconut shell: physicochemical parameters, antioxidant capacity and biological properties. J Wood Chem Technol 1–12. https://doi.org/10.1080/02773813.2024.2407374
145. Seidl R, Albrich K, Erb K, Formayer H, Leidinger D, Leitinger G, Tappeiner U, Tasser E, Rammer W (2019) What drives the future supply of regulating ecosystem services in a mountain forest landscape? For Ecol Manage 445:37–47. https://doi.org/10.1016/j.foreco.2019.03.047
146. Selvi Gökkaya D, Sert M, Sağlam M, Yüksel M, Ballice L (2020) Hydrothermal gasification of the isolated hemicellulose and sawdust of the white poplar (Populus alba L.). J Supercrit Fluids 162:104846. https://doi.org/10.1016/j.supflu.2020.104846
147. Shah MA, Khan NS, Kumar V, Qurashi A (2021) Pyrolysis of walnut shell residues in a fixed bed reactor: effects of process parameters, chemical and functional properties of bio-oil. J Environ Chem Eng 9(4):105564. https://doi.org/10.1016/J.JECE.2021.105564
148. Sharma R, Jasrotia K, Singh N, Ghosh P, Srivastava S, Sharma NR, Singh J, Kanwar R, Kumar A (2020) A comprehensive review on hydrothermal carbonization of biomass and its applications. Chem Afr 3(1):1–19. https://doi.org/10.1007/s42250-019-00098-3

149. Shen Y (2020) A review on hydrothermal carbonization of biomass and plastic wastes to energy products. Biomass Bioenergy 134:105479. https://doi.org/10.1016/j.biombioe.2020.105479
150. Siddiqui T, Khan MU, Sharma V, Gupta K (2024) Terpenoids in essential oils: chemistry, classification, and potential impact on human health and industry. Phytomedicine Plus 4(2):100549. https://doi.org/10.1016/J.PHYPLU.2024.100549
151. Soares WNC, Lira GPO, Santos CS, Dias GN, Pimenta AS, Pereira AF, Benício LDM, Rodrigues GSO, Amora SSA, Alves ND, Feijó FMC (2021) Pyroligneous acid from *Mimosa tenuiflora* and *Eucalyptus urograndis* as an antimicrobial in dairy goats. J Appl Microbiol 131(2):604–614. https://doi.org/10.1111/jam.14977
152. Sofos JN, Maga JA, Boyle DL (1988) Effect of ether extracts from condensed wood smokes on the growth of *Aeromonas hydrophila and Staphylococcus aureus*. J Food Sci 53(6):1840–1843. https://doi.org/10.1111/j.1365-2621.1988.tb07856.x
153. Souza JBG, Ré-Poppi N, Raposo JL Jr (2012) Characterization of pyroligneous acid used in agriculture by gas chromatography-mass spectrometry. J Braz Chem Soc. https://doi.org/10.1590/S0103-50532012000400005
154. Steel EA, Hinckley TM, Richards WH, D'Amore DV (2024) Forests then and now: managing for ecosystem benefits, services to humans, and healthy forests across scales. In: Future forests. Elsevier, pp 49–64
155. Stefkovics Á, Ágoston C, Bukovenszki E, Dúll A, Hortay O, Varga A (2024) Climate change worry in the times of the COVID-19 pandemic. Evidence from two large-scale European surveys. Clim Risk Manage 44:100599. https://doi.org/10.1016/J.CRM.2024.100599
156. Stott PA, Stone DA, Allen MR (2004) Human contribution to the European heatwave of 2003. Nature 432(7017):610–614. https://doi.org/10.1038/nature03089
157. Su X, Ma L, Fang Q, Yin C, Zhuang H, Qiao Y, Zhang C, Chen G (2024) Optimizing biomass combustion in a 130 t/h grate boiler: assessing gas-phase reaction models and primary air distribution strategies. Appl Therm Eng 238:122043. https://doi.org/10.1016/j.applthermaleng.2023.122043
158. Sultana R, Banik U, Nandy PK, Huda MN, Ismail M (2023) Bio-oil production from rubber seed cake via pyrolysis: process parameter optimization and physicochemical characterization. Energy Convers Manage X 20:100429. https://doi.org/10.1016/J.ECMX.2023.100429
159. Supriadi, Rahmawati S, Abram PH, Afadil, Parwati NGAM, Anggraini (2022) Characteristics of charcoal briquettes from kepok banana peel waste (Musa paradisiaca F.) as alternative fuel. Rasayan J Chem 15(01):108–115. https://doi.org/10.31788/RJC.2022.1516607
160. Suresh G, Pakdel H, Rouissi T, Brar SK, Fliss I, Roy C (2019) In vitro evaluation of antimicrobial efficacy of pyroligneous acid from softwood mixture. Biotechnol Res Innov 3(1):47–53. https://doi.org/10.1016/j.biori.2019.02.004
161. Suryani R, Anggo Rizal W, Joko Prasetyo D, Apriyana W, Anwar M, Krido Wahono S (2023) Physicochemical characteristics, antioxidant and antibacterial activities of liquid smoke derived from mixed sawdust and cocoa pod husks biomass. Trends Sci 20(6):4985. https://doi.org/10.48048/tis.2023.4985
162. Swastawati F, Muniroh M, Setiaputri AA, Sholihah DF (2023) Quality characteristics and amino acid profile and fatty acid profile of smoked catfish treated using coconut shell liquid smoke. Food Res 7(Supplementary 3):51–58. https://doi.org/10.26656/fr.2017.7(S3).8
163. Takeda S, Uchiyama J, Sugita K, Enomoto H, Ahhmed AM, Kinoshita Y, Mizunoya W, Arima Y, Sakata R (2021) Functionality of liquid smoke as an antimicrobial in cooked meat products: liquid smoke suppresses spoilage-related lactic acid bacteria. Food Sci Technol Res 27(5):759–768. https://doi.org/10.3136/fstr.27.759
164. Tavares D, Cavali M, de Tanobe VOA, Torres LAZ, Rozendo AS, Zandoná Filho A, Soccol CR, Woiciechowski AL (2022) Lignin from residual sawdust of Eucalyptus spp.—isolation, characterization, and evaluation of the antioxidant properties. Biomass 2(3):195–208. https://doi.org/10.3390/biomass2030013

165. Theapparat Y, Chandumpai A, Faroongsarng D, Theapparat Y, Chandumpai A, Faroongsarng D (2018) Physicochemistry and utilization of wood vinegar from carbonization of tropical biomass waste. Trop For New Ed. https://doi.org/10.5772/INTECHOPEN.77380
166. Theapparat Y, Khongthong S, Roekngam N, Suwandecha T, Sririyajan S, Faroongsarng D (2023) Wound healing activity: a novel benefit of pyroligneous extract derived from pyrolytic palm kernel shell wood vinegar. Ind Crops Prod 192:115994. https://doi.org/10.1016/j.indcrop.2022.115994
167. Tomczyk A, Sokołowska Z, Boguta P (2020) Biochar physicochemical properties: pyrolysis temperature and feedstock kind effects. Rev Environ Sci Biotechnol 19(1):191–215. https://doi.org/10.1007/S11157-020-09523-3/TABLES/3
168. Trilokesh C, Uppuluri KB (2019) Isolation and characterization of cellulose nanocrystals from jackfruit peel. Sci Rep 9(1):16709. https://doi.org/10.1038/s41598-019-53412-x
169. Tsai WT, Lee MK, Chang YM (2007) Fast pyrolysis of rice husk: product yields and compositions. Bioresour Technol 98(1):22–28. https://doi.org/10.1016/J.BIORTECH.2005.12.005
170. Ucella-Filho JGM, Ferreira NS, Alves MR, Ignacchiti MDC, Dias Júnior AF, Resende JA (2024) Evaluation of natural products as therapeutic alternatives for bovine mastitis and implications for future research. S Afr J Bot 167:310–321. https://doi.org/10.1016/j.sajb.2024.02.031
171. Ucella-Filho JGM, Ferreira NS, Cupertino GFM, Guedes NA, de Paula H, Kitagawa RR, Balu AM, Luque R, Ignacchiti MDC, Parreira LA, Resende JA, Dias Junior AF (2024) Antiviral potential of bio-oil from Citrus sinensis waste wood as a therapeutic approach against COVID-19. J Clean Prod 447:141583. https://doi.org/10.1016/J.JCLEPRO.2024.141583
172. Ucella-Filho JGM, Ferreira NS, Lacerda EMF, Soriano Neto C, de Duarte CES, Ferreira IM, Balu AM, Ignacchiti MDC, Dias Junior AF, Resende JA (2024) Citrus sinensis waste wood pyroligneous extract as an effective agent against biofilms formed by mastitis-causing Staphylococcus aureus. Ind Crops Prod 218:118854. https://doi.org/10.1016/j.indcrop.2024.118854
173. Ucella-Filho JGM, da Freire ASM, Carréra JC, Lucas FMF, Zucolotto SM, Dias Júnior AF, Mori FA (2022) Tannin-rich bark extract of plants as a source of antimicrobial bioactive compounds: a bibliometric analysis. S Afr J Bot 150:1038–1050. https://doi.org/10.1016/J.SAJB.2022.09.018
174. Ucella-Filho JGM, Lorenço MS, da Silva BRF, Ferreira VRF, Monteiro JRB, Ferreira NS, das Cardoso MG, Kitagawa RR, Resende JA, Dias Junior AF, Piccoli RH, Mori FA (2024) Exploring the potential of tannin-rich tree bark extracts in combating foodborne diseases and gastric cancer. Food Biosci 57:103559. https://doi.org/10.1016/J.FBIO.2023.103559
175. Umriani Permatasari N, Dali S, Sry Wahyuni E (2023) Potential Cassava Peel Waste (Manihot esculenta Crantz) in the production of bioethanol by enzymatic hydrolysis and fermentation using Zymomonas mobilis bacteria. J Pure Appl Chem Res 12(2):87–103. https://doi.org/10.21776/ub.jpacr.2023.012.02.3304
176. Varma AK, Thakur LS, Shankar R, Mondal P (2019) Pyrolysis of wood sawdust: effects of process parameters on products yield and characterization of products. Waste Manage 89:224–235. https://doi.org/10.1016/J.WASMAN.2019.04.016
177. Velmurugan N, Chun SS, Han SS, Lee YS (2009) Characterization of chikusaku-eki and mokusaku-eki and its inhibitory effect on sapstaining fungal growth in laboratory scale. Int J Environ Sci Technol 6(1):13–22. https://doi.org/10.1007/BF03326056
178. Vieira WT, Bispo MD, Amaral GVM, de Farias MB, de Melo SF, de Almeida ADSV, de Carvalho SHV, de Oliveira Pimentel WR, Vieira MGA, Dos Santos Araújo MA, Soletti JI, Balliano TL (2023) Evaluation of the use of pyroligneous extract obtained from vacuum pyrolysis of residual biomass for onychomycosis treatment. Braz J Chem Eng. https://doi.org/10.1007/s43153-023-00376-1
179. Villamor GB, Wallace L (2024) Corporate social responsibility: current state and future opportunities in the forest sector. Corp Soc Responsib Environ Manage 31(4):3194–3209. https://doi.org/10.1002/csr.2743

180. Vuppaladadiyam AK, Vuppaladadiyam SSV, Awasthi A, Sahoo A, Rehman S, Pant KK, Murugavelh S, Huang Q, Anthony E, Fennel P, Bhattacharya S, Leu S-Y (2022) Biomass pyrolysis: a review on recent advancements and green hydrogen production. Bioresour Technol 364:128087. https://doi.org/10.1016/j.biortech.2022.128087
181. Wang C, Mei J, Zhang L (2021) High-added-value biomass-derived composites by chemically coupling post-consumer plastics with agricultural and forestry wastes. J Clean Prod 284:124768. https://doi.org/10.1016/j.jclepro.2020.124768
182. Wang S, Dai G, Yang H, Luo Z (2017) Lignocellulosic biomass pyrolysis mechanism: a state-of-the-art review. Prog Energy Combust Sci 62:33–86. https://doi.org/10.1016/j.pecs.2017.05.004
183. Waters CL, Janupala RR, Mallinson RG, Lobban LL (2017) Staged thermal fractionation for segregation of lignin and cellulose pyrolysis products: an experimental study of residence time and temperature effects. J Anal Appl Pyrol 126:380–389. https://doi.org/10.1016/J.JAAP.2017.05.008
184. Wei Q, Ma X, Dong J (2010) Preparation, chemical constituents and antimicrobial activity of pyroligneous acids from walnut tree branches. J Anal Appl Pyrol 87(1):24–28. https://doi.org/10.1016/j.jaap.2009.09.006
185. Wu S, Wang C, Chen D, Zhou W, Chen X, Wang M, Zhang Q (2022) Effects of pyroligneous acid as silage additive on fermentation quality and bacterial community structure of waste sugarcane tops. Chem Biol Technol Agric 9(1):67. https://doi.org/10.1186/s40538-022-00335-x
186. Xin X, Zhao W, Essien S, Dell K, Baroutian S (2022) The effects of ageing treatment on bioactive contents and chemical composition of liquid smoke food flavourings. Eur Food Res Technol 248(5):1311–1319. https://doi.org/10.1007/s00217-022-03976-2
187. Xue R, Cui E-L, Hu G-Q, Zhu M-Q (2022) The composition, physicochemical properties, antimicrobial and antioxidant activity of wood vinegar prepared by pyrolysis of Eucommia ulmoides Oliver branches under different refining methods and storage conditions. Ind Crops Prod 178:114586. https://doi.org/10.1016/j.indcrop.2022.114586
188. Yadav K, Jagadevan S, Yadav K, Jagadevan S (2019) Influence of process parameters on synthesis of biochar by pyrolysis of biomass: an alternative source of energy. Recent Adv Pyrol. https://doi.org/10.5772/INTECHOPEN.88204
189. Yao S, Nie S, Yuan Y, Wang S, Qin C (2015) Efficient extraction of bagasse hemicelluloses and characterization of solid remainder. Bioresour Technol 185:21–27. https://doi.org/10.1016/j.biortech.2015.02.052
190. Yek PNY, Wan Mahari WA, Kong SH, Foong SY, Peng W, Ting H, Liew RK, Xia C, Sonne C, Tabatabaei M, Almomani F, Aghbashlo M, Lam SS (2022) Pilot-scale co-processing of lignocellulosic biomass, algae, shellfish waste via thermochemical approach: recent progress and future directions. Bioresour Technol 347:126687. https://doi.org/10.1016/j.biortech.2022.126687
191. Yin D, Xue R, Li Y, Zhu M, Li D (2024) Valorization of Coptis chinensis extraction residue via slow pyrolysis for the production of bioactive wood vinegar. Biomass Convers Biorefin 14(14):16559–16574. https://doi.org/10.1007/s13399-023-03890-x
192. Yu Y, Deng Z, Li C, He X, Chang J (2024) Enhanceing antifungal activity of starch-based wood adhesive by Larch bio-oil. Int J Adhes Adhes 130:103612. https://doi.org/10.1016/j.ijadhadh.2023.103612
193. Yupa NP, Purwaningsih H, Kemala T (2023) Synthesis and characterization of corn husk (Zea Mays L.) cellulose using microwave-assisted extraction (MAE). Jurnal Kimia Sains dan Aplikasi 26(7):268–275. https://doi.org/10.14710/jksa.26.7.268-275
194. Zeleke NM, Sinha DK, Mengesha GA (2022) Chemical composition and extraction of micro crystalline cellulose from outer skin isolated coffee husk. Adv Mater Sci Eng 2022:1–13. https://doi.org/10.1155/2022/7163359
195. Zhang L, Dou X, Yang Z, Yang X, Guo X (2021) Advance in hydrothermal bio-oil preparation from lignocellulose: effect of raw materials and their tissue structures. Biomass 1(2):74–93. https://doi.org/10.3390/biomass1020006

196. Zhang Q, Chang J, Wang T, Xu Y (2007) Review of biomass pyrolysis oil properties and upgrading research. Energy Convers Manage 48(1):87–92. https://doi.org/10.1016/j.enconman.2006.05.010
197. Zhong X, Ashiru-Oredope D, Pate A, Martin GP, Sharma A, Dark P, Felton T, Lake C, MacKenna B, Mehrkar A, Bacon SCJ, Massey J, Inglesby P, Goldacre B, Walker AJ, Curtis HJ, Morley J, Hickman G, Croker R, Evans D, Ward T, DeVito NJ, Fisher L, Green ACA, Smith RM, Hulme WJ, Davy S, Andrews CD, Hopcroft LEM, Dillingham I, Higgins R, Cunningham C, Wiedemann M, Nab L, Maude S, Macdonald O, Butler-Cole BFC, O'Dwyer T, Stables CL, Wood C, Brown AD, Speed V, Bridges L, Schaffer AL, Walters CE, Bates C, Cockburn J, Parry J, Hester F, Harper S, Hand K, Bladon S, Cunningham N, Gilham E, Brown CS, Mirfenderesky M, Palin V, van Staa TP (2023) Clinical and health inequality risk factors for non-COVID-related sepsis during the global COVID-19 pandemic: a national case-control and cohort study. EClinicalMedicine 66:102321. https://doi.org/10.1016/J.ECLINM.2023.102321
198. Zhou K, Midkiff D, Yin R, Zhang H (2024) Carbon finance and funding for forest sector climate solutions: a review and synthesis of the principles, policies, and practices. Front Environ Sci 12. https://doi.org/10.3389/fenvs.2024.1309885

Biochar and Bonechar Produced from Agro-industrial Waste in Water Treatment and Immobilization of Residual Herbicides in the Soil

Fernanda Aparecida Nazário de Carvalho, Iara Nobre Carmona, Merlindo Jacinto Manjate, Juliana Terezinha Santos Cruz, Allana Katiussya Silva Pereira, Rodrigo Nogueira de Sousa, Kassio Ferreira Mendes, and Ananias Francisco Dias Júnior

Abstract Agro-industries generate large amounts of waste which, when improperly disposed of, negatively impact the environment and public health. The improper disposal of wood waste, sugarcane bagasse, animal bone waste from the meat

---

F. A. Nazário de Carvalho (✉) · I. N. Carmona · M. J. Manjate · J. T. S. Cruz · A. F. Dias Júnior
Agricultural Sciences and Engineering Center, Universidade Federal do Espírito Santo (Ufes), Av. Governador Lindemberg, 316, Jerônimo Monteiro, ES 29550-000, Brazil
e-mail: fernandacarvalhonaz@gmail.com

I. N. Carmona
e-mail: iaranobrecarmona@gmail.com

M. J. Manjate
e-mail: manjatemerlindo@gmail.com

J. T. S. Cruz
e-mail: santosjuliana869@gmail.com

A. F. Dias Júnior
e-mail: ananias.dias@ufes.br

I. N. Carmona · A. K. S. Pereira
Department of Forest Sciences, "Luiz de Queiroz" College of Agriculture (Esalq), Universidade de São Paulo (USP), Av. Pádua Dias, 11, Piracicaba, SP 13418-900, Brazil
e-mail: allana.florestal@gmail.com

M. J. Manjate
Faculty of Agricultural Sciences, Universidade Lúrio (Unilúrio), Unango Campus, CEP 3300 Sanga, Niassa, Moçambique

R. N. de Sousa · K. F. Mendes
Center for Nuclear Energy in Agriculture (Cena), Universidade de São Paulo (USP), Av. Centenário, 303, Piracicaba, SP 13400-970, Brazil
e-mail: rodrigosouza@usp.br

K. F. Mendes
e-mail: kfmendes@cena.usp.br

A. K. Silva Pereira and A. F. Dias Júnior (eds.), *Agro-industrial Waste for New Pharmaceuticals*, Green Energy and Technology,
https://doi.org/10.1007/978-3-031-88534-1_3

industry, and even residues from grain production industries, through burning or inappropriate dumping, can release toxic gases, contribute to greenhouse gas emissions, and facilitate the proliferation of pathogens. Thus, the objective of this chapter was to discuss the key variables related to the improper management of agro-industrial waste and its environmental consequences, highlighting bioproducts obtained through the pyrolysis process, such as biochar and bonechar, as potential solutions. To this end, the chapter addresses the impacts of population growth on increased agro-industrial production and waste generation, the problems associated with the improper disposal of these byproducts—such as landfill deposition and burning—and the potential of biochar and bonechar, which not only enable the sustainable reuse of these materials but also serve as materials capable of remediating contaminated soils, water sources, and effluents. Consequently, it can be observed that using these residues for the production of sustainable materials minimizes environmental impacts, reduces the consumption of non-renewable resources, and promotes a circular economy, contributing to sustainable development and ecosystem preservation. Furthermore, the pyrolysis process that produces these bioproducts requires the adoption of low-cost technologies, promoting industrial symbiosis, reducing environmental impact, and creating economically viable opportunities, aligning with sustainable development goals and contributing to the reduction of waste generated by agro-industries.

Keywords Industrial symbiosis · Carbonaceous products · Environmental management

1 Introduction

Although there is an abundance of information related to food preparation from a gastronomic perspective, there is a noticeable lack of technological actions focused on environmentally responsible disposal practices for the by-products generated by the food industries—agro-industries. Population growth has driven a significant transformation in consumption patterns, promoting the globalization of trade. Today, the global population is three times larger than it was in the mid-twentieth century, and it is projected to reach 8.5 billion in 2030 and 9.7 billion in 2050 [164]. Among consumption sectors, the agro-industrial field faces some of the greatest challenges due to population growth. High production demand brings logistical challenges, particularly regarding waste generation, as greater production leads to higher amounts of by-products.

Agro-industrial waste is considered an undesirable by-product of the production of primary raw materials. In this context, it is estimated that approximately 1.3 billion tons of agro-food waste are generated annually, equivalent to one-third of all food production [49]. Examples include waste from meat processing, crop harvesting, and food consumption [133]. These by-products encompass various materials, such as slaughterhouse and meat processing waste, manure or animal waste, field crop residues, including harvest remains, husks, straws, stems, and peels of fruits and vegetables, which pose a significant challenge regarding improper management and the negative impacts on soil, water, and biodiversity [78].

The improper disposal of agro-industrial waste in the environment intensifies pressure on ecosystems, exacerbating pollution and contributing to ecological imbalances, in addition to generating serious consequences for public health. A large portion of these wastes is not treated or reused, often being discarded through burning, improper dumping, or uncontrolled landfilling [127]. This disposal causes serious environmental, economic, and social problems, as well as promoting the proliferation of insects, rodents, birds, and microorganisms. Although these organisms are essential for decomposition, in excess, they can become harmful, contributing to the emission of greenhouse gases, the release of toxic gases, and the proliferation of pathogenic bacteria and fungi [72]. Therefore, an alternative that should be enabled is the increased use of biomass resulting from agro-production processes, with the potential to make significant contributions to mitigating the damage [58].

The United Nations [166] highlighted in its global action plan (Agenda 2030) that to achieve Sustainable Development, it is necessary to ensure sustainable consumption and production patterns. Specifically regarding the reduction of waste generation, by adopting circular economy strategies such as prevention, reduction, recycling, and reuse. An interesting approach to dealing with these wastes is to promote their utilization, either by transforming them into bioactive compounds as alternatives to synthetic medicines [162], or into biohydrogen, bio-oils, biogas, and even biochar.

Biochar is a valuable, inexpensive, and sustainable resource that finds applications in multidimensional spheres. The numerous advantages of biochar highlight its contribution to the energy sector, due to its specific chemical composition, depending on the source material [141]. Over the years, biochar has also proven to be a material capable of removing organic and inorganic pollutants (herbicides, fungicides, insecticides, heavy metals, antibiotics, industrial chemicals, among others) from soil and water through sorption–desorption and degradation processes [20]. The residual presence of herbicides in soil is one of the problems that biochar can help mitigate efficiently. Herbicides that remain in the soil for long periods can cause environmental contamination, affect beneficial organisms, and harm the development of subsequent crops. Thanks to its high surface area, porosity, and chemical functional groups, biochar has an excellent sorption capacity, reducing the mobility and bioavailability of these compounds. Furthermore, biochar can act as a physical barrier, limiting the leaching of herbicides into groundwater and minimizing the risks of water pollution. One example is the herbicide fomesafen, whose residual effect can cause toxicity to wheat in crop rotation, reducing its productivity. Biochar, made from wheat straw, was tested at three concentrations (1, 2, and 4%) to evaluate its effects on wheat growth, plant health, and the microbial community in soil contaminated with fomesafen. The results showed that biochar reduced herbicide absorption by wheat, eliminating its toxicity. Additionally, it improved the presence of beneficial bacteria and fungi in the wheat rhizosphere, with the 2% biochar concentration being the most effective in reducing toxicity and balancing the microbial community in the contaminated soil [107].

Biochar is a carbonaceous material produced by the thermochemical conversion of biomass in a controlled or oxygen-free environment and can be named according to its biomass source [178]. It can be found under different names, such as bonechar, which gets its name from being produced from the pyrolysis of animal bones, often derived from slaughterhouse waste [182]. Thus, bonechar is a type of biochar that exhibits high porosity and has the potential to act as an adsorbent and even a fertilizer when added to soils. However, unlike biochar, which is predominantly carbon, bonechar is composed of tricalcium phosphate (hydroxyapatite) and contains low total carbon content, making it a unique material deserving special study [103].

By utilizing biomass waste from agroindustries, a sustainable replacement of non-renewable raw materials is promoted, and the costs of waste disposal in landfills are eliminated [88]. Therefore, finding alternatives to reduce damage to natural resources has become an essential task for researchers, aiming to protect the interests of current and future generations [53]. This opportunity aligns with the interests outlined in the sixth assessment report of the Intergovernmental Panel on Climate Change (2023), which addresses the importance of changing production, consumption, and disposal patterns to minimize the effects of climate change. Additionally, in order to reduce emissions, the world needs to consider a material transition, generating new materials with low impact, promoting the reuse of industrial process byproducts [88]. This way, it is possible to reduce energy consumption and minimize environmental pollution, promoting industrial symbiosis. This reduces environmental impacts and operational costs for companies by utilizing waste with low-cost technologies, while also creating an alternative income source.

Thus, the objective of this chapter is to present the technological advances for the reuse of agroindustrial waste, focusing on the production of biochar and bonechar as sustainable solutions to mitigate environmental impacts. The text will explore the production of agroindustrial waste and the challenges related to the improper management of these wastes, discussing the feasibility of using biochar and bonechar for the sorption of contaminants and the improvement of soil and water quality. Furthermore, it will address the contribution of these practices to the circular economy, climate change mitigation, and the development of new materials with low environmental impact, highlighting the integration of these technologies as a crucial strategy to promote sustainability in production processes and the protection of ecosystems.

2 Agroindustrial Activities that Generate Waste Which Can Be Valorized Through the Production of Biochar and Bonechar

The current perspective for the development of sustainable industries involves the gradual replacement of non-renewable sources with renewable ones [52]. This factor can be made feasible through agricultural and industrial activities that generate, continuously and on a large scale, large volumes of waste. These wastes, when not properly treated, cause serious environmental problems that affect various compartments of the ecosystem, such as air, water, and soil [5, 36]. According to Sagar et al. [134], air pollution is increased with the burning or improper disposal of waste, releasing harmful pollutants that affect air quality and public health, contributing to the intensification of the greenhouse effect. Similarly, water contamination is aggravated by leachate runoff, which causes eutrophication and creates dead zones in water bodies. This highlights the need to promote more beneficial disposal of these wastes.

One of the sectors generating waste is the sugarcane industry. Sugarcane is one of the essential strategic agricultural products, with a global production of sugarcane exceeding 75 million tons annually [92, 124]. Meanwhile, Brazil's sugarcane production in the 2023/2024 harvest reached 713.2 million tons, setting a new record in the historical series monitored by the National Supply Company. According to CONAB, the biomass produced by sugarcane is widely used in the production of two products that are essential to the global economy: sugar, accounting for about 35% of Brazil's total production, and ethanol, which accounts for about 65%. From this production, some waste is generated, with sugarcane bagasse being the main byproduct. It is estimated that between 5 and 12 million tons of bagasse remain annually, representing about 30% of the milled sugarcane. According to the Federation of Rural Associations of the State of São Paulo [46], sugar mills have been using 60–90% of this bagasse as an energy source for electricity generation, in the paper and cardboard industry, in the manufacture of composites, in the chemical industry, and in animal feed production. Despite the various uses of sugarcane bagasse, there is still a surplus of 10–30% in mills and distilleries. This surplus is often disposed of inadequately, as in the case of straw and bagasse, which, although organic materials, degrade quickly when left exposed to the soil, increasing the release of gases such as carbon dioxide (CO_2), methane (CH_4), and nitrous oxide (N_2O) [155, 156]. Mielke et al. [109] highlighted the efficiency of biochar produced from this raw material at a pyrolysis temperature of 750 °C for the immobilization of the herbicide metribuzin in Oxisol, suggesting a potential use for this bioproduct.

Another important crop is peanuts (Arachis hypogaea), which is one of the main oilseeds in the world. The peanut seed contains approximately 40–45% oil and 25% protein and is widely grown for consumption as a snack after roasting or for oil extraction used in cooking [96]. However, it generates some byproducts in the process, such as shells and press cake, which are commonly used as animal feed and also for the production of peanut flour [19]. According to the same author, peanuts are the third largest oilseed produced globally and the second largest in India. India holds the second position in terms of production and the first in terms of area, while China is the largest producer and consumer of peanuts in the world. In the last decade, peanut production in Brazil has grown significantly, accompanied by an increase in the cultivated area. According to CONAB [23], the 2014/15 harvest increased from 346.8 thousand tons to an estimated 746.7 thousand tons in the 2021/22 cycle, a 115% increase. According to Abdelmagid and Qiu [1], one of the main components of peanut production is the shell, which constitutes ~30% of the total mass of peanuts and generates ~5.4 million metric tons of waste. These authors further emphasized that a portion of the shells produced is discarded or burned, leading to resource depletion and environmental pollution. Kayranli et al. [79] found the efficiency of peanut shell biochar for removing rhodamine B, a basic dye used in the textile industry, which generates highly colored wastewater that is harmful to the environment due to its toxicity and mutagenicity.

Another highly relevant crop is rice (*Oryza sativa*), one of the most cultivated crops in the world and a staple food for more than half of the global population. With a total production of 756.74 million tons per year, Asian countries account for more than 90% of this production, with China being the largest producer, followed by India [45]. The development of high-yielding, short-cycle varieties has consolidated rice as one of the pillars of the Asian agricultural system [145]. In Brazil, the 2023/24 rice harvest had a planted area of approximately 1606.9 thousand hectares, with a productivity of 6587 kg/ha and a production of 10,585.3 thousand tons, despite the negative influence of climatic adversities on rice crop productivity. With such high rice productivity, the byproducts generated in production must also be a cause for concern. According to Bhattacharyya et al. [22], an equivalent of 731 million tons of rice straw is produced globally, with India contributing about 126.6 million tons, of which 60% is burned in the field due to technical, structural, institutional, and socio-economic challenges in utilizing it. As a result, only 20% of the rice straw produced worldwide is utilized, and more than 100 million tons of rice straw are burned annually around the world as an alternative disposal method for this waste [145]. Therefore, it is necessary to find viable alternatives for utilizing these "undesirable" inputs, aiming for environmentally correct and economically feasible solutions, transforming these wastes into raw materials that generate value for rural communities. Sakhiya et al. [136] suggested that the use of rice husk as biochar is highly efficient in improving water quality in areas contaminated with arsenic and manganese. The biochar produced through slow pyrolysis at 600 °C had a production cost of US$ 0.913/kg, making it an economical adsorbent. Groundwater treated with 0.1 g of biochar showed high removal efficiency (> 85%), reducing the environmental contamination of water, which is essential for sustaining life.

Coffee (Coffea spp.) is a widely appreciated product in many countries around the world, known for its rich variety of aromas and flavors that delight the palates of millions of people. This spice can be grown in different regions of the world, from the mountains of Latin America to the plains of Africa and Asia, offering a diversity of sensory profiles influenced by factors such as climate, soil, and cultivation methods. Coffee plays a culturally significant role in many societies and is often associated with moments of socialization and interaction, transcending borders in its popularity, making it one of the most consumed beverages globally. But does the high production of coffee, due to its great appreciation worldwide, generate any waste? Annually, the coffee industry generates more than 10 million tons of coffee waste worldwide [42]. The husks are the main waste produced during coffee production and can account for up to 50% of the weight of dried beans, yet they still lack a well-established application, often being improperly discarded or burned [140]. This waste is a response to the high global consumption of this product, which, in 2022 alone, reached about 10.7 billion kilograms consumed, marking a 4.2% increase from the previous year [83]. According to the USDA [161], Brazil is the world's largest coffee producer, accounting for 39% of global production, having produced 63.6 million 60-kg bags in the 2023/24 harvest. Following Brazil, but with much lower averages, are Vietnam with a significant 17%, Colombia with 7%, Ethiopia and Indonesia with 5%, and other countries together accounting for 11% of the total production. Given the high productivity, sustainable alternatives can be devised to enable economic and environmental gains from these byproducts. Benedicto et al. [34] highlighted the success of biochars produced from coffee husks as an easy and cheap alternative for surfactant adsorption. Studies like this serve as an incentive to promote new uses for plant-based agro-industrial waste.

Animal-based raw materials are also important inputs in the chain of products generated by agroindustries. The meat processing industry is one of the main sectors responsible for the processing of beef, pork, and poultry. In Brazil, during the first quarter of 2024 alone, 9.237 million head of cattle, 13.939 million head of pigs, and 1.591,653 billion chickens were slaughtered [70]. But does this high slaughter demand produce waste? The answer is quite obvious when the data is analyzed and calculated. Each pig slaughtered, based on data from IBGE, produced an average of approximately 91.77 kg of carcass, while each chicken produced about 2.11 kg of carcass, and for cattle, approximately 258 kg of carcass per animal. The nature and quantity of the waste vary at each stage, but they include carcasses of dead animals, parts of animals considered inedible, bones, skins, and blood. The most common slaughterhouse solid waste is animal bone, where, of the total waste, cattle contribute about 46%, pigs 38%, and poultry 28% [158]. Approximately 0.13 billion tons of animal bone waste are generated annually by the meat production industry worldwide [69]. It is known that the improper disposal of these highly perishable wastes is responsible for polluting the environment by favoring the reproduction system of pathogenic microorganisms, which become an additional health risk, as well as being a major source of greenhouse gas emissions. Some authors suggest the use of these bones in carbonized form, known as bonechar. Bonechar is a biochar with a low fixed carbon content but rich in calcium phosphate, the main component of human and animal bones, composed of approximately 10% carbon and 90% hydroxyapatite [125]. Due to its composition, this material has been gradually explored as a fertilizer because of its phosphorus richness and also for sorption purposes, such as fluoride, demonstrating its superior capacity compared to other non-renewable sorbents in water sources [65, 103, 125, 126].

In this context, understanding the numerous wastes generated by agroindustries is essential to enable alternatives that add value to these various byproducts, creating alternative income for the producing industry, as well as environmental benefits. It is known that the global agri-food industry generates billions of tons of byproducts annually; however, these byproducts are unfortunately often discarded as waste, becoming environmental pollutants that exacerbate the greenhouse effect and leachate, which may reach groundwater (Fig. 1) [57]. Thus, aiming to valorize animal and plant biomass, transforming it into solutions for agriculture—which is the most important economic activity in much of the world—it is necessary to support the creation of sustainable bioproducts that can be applied as solutions for different

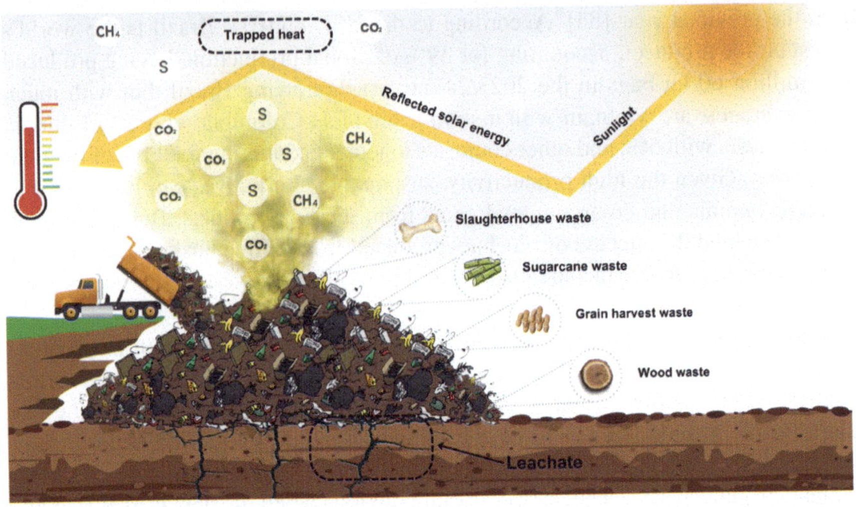

Fig. 1 Diagram of improper disposal of agroindustrial waste in landfills and its environmental consequences. *Source* Authors (2024)

spheres of the ecosystem. Proposals like these align directly with the United Nations Sustainable Development Goals (SDGs), contributing to the promotion of sustainable use of terrestrial ecosystems (Goal 15), combating climate change and its impacts (Goal 13), and ensuring sustainable production and consumption (Goal 12). Thus, it is possible to strengthen ties, creating a network of work, support, and partnerships for sustainable development (Goal 17), promoting innovation and fostering new ideas for human and global well-being.

3 Transformation of Agroindustrial Waste into Biochar and Bonechar

Environmental concerns arising from climate change and its potential negative consequences require the implementation of effective solutions for recycling large volumes of waste continuously generated by the meat processing industries [9]. The growing interest in the use of biochars, such as bonechar, hydrochar, lignochar, among others, has driven the conversion of biomass into sustainable solutions that make use of waste. Hydrochar and biochar are carbonaceous materials widely used in agriculture to improve soil quality and in contamination remediation programs, due to their economic feasibility and applicability [81, 154]. The main difference between hydrochar and biochar lies in the thermal treatment conditions. Hydrochar is obtained through hydrothermal carbonization, a thermochemical process that uses high moisture to convert biomass into a carbonaceous material [122, 154], using temperatures from 180 to 230 °C under high pressure [50].

On the other hand, biochar is a carbon-rich product obtained when biomass is subjected to thermal treatment in an oxygen-limited environment [76, 89, 154], known as pyrolysis, which typically occurs at temperatures between 300 and 900 °C [170]. The process results in the formation of carbon-rich solid products (biochar), a volatile fraction that can be partially condensed into a liquid phase (bio-oil), and the so-called "non-condensable" gases, such as CO, CO_2, CH_4, and H_2 [97, 170]. This process requires pre-treatment for drying the wet biomass and a higher energy demand. Thus, hydrochar, due to its production method, offers advantages in terms of performance and economic efficiency compared to biochar [77]. Bonechar is produced using the same process as biochar, typically between 500 and 1000 °C,however, in addition to being a carbon-rich product, carbonization preserves its mineral structure, mainly hydroxyapatite $Ca_{10}(PO_4)_6(OH)_2$ [7, 6]. The raw materials used for the production of biochar, as well as hydrochar, include lignocellulosic material [77, 129]. Among these organic raw materials are agricultural, forest, and urban wastes [7, 6, 95], and their properties can vary depending on the raw material used and the pyrolysis conditions of production [6–6]. On the other hand, bonechar is specifically derived from animal bones, globally used from cattle, sheep, chicken, and fish [7, 6, 100], and is known for its unique composition, which includes a high content of calcium phosphate, giving it distinct properties and potential for applications in areas such as environmental remediation and sustainable agriculture [125, 126].

3.1 Characteristics of Biochar and Bonechar

Biochar has received increasing attention due to its unique characteristics, such as high carbon content and cation exchange capacity (CEC), large specific surface area (SSA), stable structure, and strong adsorption capacity. The main characteristics often used to assess the quality of biochar include carbon content, apparent density, carbon stability, volatile compound content, surface area, nutritional content, heavy metals, pH value, water and ash content, polycyclic aromatic hydrocarbons (PAHs), water retention capacity, porosity, elemental composition, cation exchange capacity, and electrical conductivity [80, 116, 138].

The yield and quality of biochar obtained from thermo-chemical biomass conversion processes are strongly influenced by the feedstock and pyrolysis conditions [171]. These factors vary widely due to differences in available oxygen, heating rate, and reaction temperature. Parameters such as heating rate, residence time, and temperature during pyrolysis play a crucial role in defining the final characteristics of biochar [10, 75, 160]. These variations in parameters can lead to products with a wide range of pH values, specific surface area, pore volume, CEC, volatile matter, ash content, and carbon content [159], as well as total nitrogen, potassium, phosphorus, calcium, magnesium, sodium, and sulfur content [80].

Different types of biochar can present distinct characteristics, even when produced from various feedstocks under the same pyrolysis conditions. The properties of biochar are strongly influenced by the biomass used in its production. The physical properties of biochar, such as density and porosity, are influenced by pyrolysis temperature. In general, higher temperatures (600–800 °C) tend to increase the density of biochar and modify its porous structure [152], which can impact its ability to retain water and nutrients [80]. The type of feedstock significantly influences the apparent density of biochar. Khater et al. [80] observed that increasing the pyrolysis temperature from 400 to 800 °C resulted in an increase in the apparent density of the studied materials: from 161.5 to 187.1 kg m^{-3} for rice straw, from 195.0 to 211.9 kg m^{-3} for sawdust, from 175.7 to 194.1 kg m^{-3} for sugarcane, and from 188.0 to 199.4 kg m^{-3} for tree leaves. The highest apparent density, 211.9 kg m^{-3}, was observed in sawdust at 800 °C, while the lowest apparent density, 161.5 kg m^{-3}, was recorded in rice straw at 400 °C. Biomasses with higher lignin content tend to produce biochar with higher density due to their more complex structure, which is more resistant to decomposition during pyrolysis [152].

SSA is a very important characteristic that influences its properties and applications, especially in the adsorption of nutrients and pollutants. As the pyrolysis temperature increases, the SSA also tends to increase [71, 149, 152, 159], and the pore volume increases significantly with the rise in temperature, indicating that higher temperatures result in more porous biochars with greater nutrient and water retention capacity [71]. Ahmad et al. [2] observed that at 300 °C, biochar derived from soybean straw had a surface area of 6 m^2/g, while the one derived from peanut shells had 3 m^2/g. When the temperature was increased to 700 °C, the surface area of soybean straw increased to 420 m^2/g, and the peanut shell biochar reached 448 m^2/g. This change affected the overall porosity and surface area of the biochar, essential factors that influence nutrient retention and availability in the soil [153]. This characteristic makes the material effective in adsorbing organic and inorganic contaminants due to its high surface area and the development of micropores [163].

The pyrolysis temperature also affects the yield and volatile material content. As the pyrolysis temperature increases, both biochar yields and volatile matter content gradually decrease [26, 82, 151, 160], while the production of gaseous and liquid products increases [26]. Sun et al. [151] observed that at lower temperatures (e.g., 300 °C), biochar retains more volatile compounds, whereas at higher temperatures (e.g., 600 °C), biochar becomes richer in carbon and more stable. The decrease in volatile material content is explained by the increased aromatization and greater losses of gaseous products, tar, and low molecular weight hydrocarbons resulting from the higher pyrolysis temperature [37]. In addition to temperature, residence time is a crucial parameter in biochar production through pyrolysis. It varies with pyrolysis temperature [151], heating rate, and other parameters, making it very challenging to provide a clear idea of the role of residence time in biochar production [160]. Sakhiya et al. [135] reported that biochar yield decreased by up to 10% with an increase in residence time from 1 to 3 h at 300 °C, as more volatile components are released with more time, resulting in greater formation of bio-oil and gases.

Residence time not only affects the yield of biochar but also influences its quality and characteristics, promoting the development of micro and macropores [160]. The choice of residence time should be based on the desired characteristics of the biochar for specific applications. At lower temperatures, a longer residence time may result in the removal of irregular carbon and the opening of micropores, increasing the surface area of the biochar, which can enhance its adsorption and nutrient retention properties [151].

In addition to temperature, the feedstock also has a strong influence on the composition of volatile matter. Jindo et al. [75] observed that wood-derived biochars, such as apple and oak branches, showed greater variation in volatile content at pyrolysis temperatures between 400 and 800 °C, compared to biochars from non-woody materials, due to the higher resistance of lignin to decomposition at lower temperatures. Ash content also increases with pyrolysis temperature. There is a strong positive correlation between pyrolysis temperature and ash content, as the pyrolysis temperature increases, most of the inorganic components present in the biomass condense and are retained in the biochar, leading to an increase in ash content [179].

The pH of biochar is influenced by both the pyrolysis temperature and the type of feedstock used. pH tends to increase with increasing pyrolysis temperature [80, 132]. Zhang et al. [180] reported that biochar produced from herbs and woody plants tends to decompose acidic functional groups, such as carboxyl and phenolic hydroxyl, and volatilize organic acids during pyrolysis at high temperatures, resulting in pH generally between 7 and 10.4. In contrast, biochar derived from animal manure, due to its high content of mineral oxides and carbonates, has a pH ranging from 7.5 to 11.4, and for sewage sludge, pH typically ranges from 8.46 to 11.7. The increase in temperature and, consequently, in pH is attributed to the higher concentration of non-pyrolyzed elements present in the feedstock [80, 116]. Ferjani et al. [71] observed that the increase in temperature favored the increase in neutral pH around 7.2 at 300 °C to more alkaline values up to 10.1 at 600 °C, and this change in pH may influence nutrient availability and soil chemistry, making biochar more suitable for agricultural applications.

Like pH, electrical conductivity (EC) is another characteristic influenced by temperature. As pyrolysis temperature increases, EC also increases [80, 151]. Khater et al. [80] observed that EC varied with temperature in different types of feedstocks. Tree leaves exhibited high values of 3.46 dS m^{-1} at 800 °C, compared to rice straw, sawdust, and sugarcane residues. Cation exchange capacity (CEC) tends to decrease with increasing pyrolysis temperature [37, 71, 159]. This mainly occurs due to the removal of surface functional groups, such as carboxyl and phenolic groups, which are responsible for cation exchange [76, 149].

Fixed carbon content varies with pyrolysis temperature and feedstock type. Increasing temperature leads to an increase in fixed carbon [99] and a decrease in hydrogen and oxygen content [80, 174], due to the removal of functional groups containing these elements during pyrolysis [174]. Huang et al. [68] reported that biochar produced at 600 °C from wood had a high carbon content (83.29%), while chicken manure biochar at 300 °C had a very low carbon content of 32.46%. However, in biochars derived from animal waste, such as chicken manure, carbon concentration can decrease as temperature increases, suggesting that the organic compounds in these materials are more unstable and degrade rapidly at high temperatures [37]. Volatile matter content in biochar decreases with increasing pyrolysis temperature, while ash content tends to increase. Sun et al. [151] observed that at 300 °C, volatile matter content ranged from 34 to 54%, but it significantly decreased to 6–19% when the temperature reached 600 °C. In contrast, ash content increased, with biochar derived from forest litter showing the highest ash content among the different types of biomass evaluated. Pyrolysis temperature has a significant impact on the nutrients in biochar. Total nitrogen content decreases with increasing temperature, resulting from the breakdown of nitrogen-containing compounds [149, 179]. In contrast, micronutrient concentrations, such as calcium, iron, magnesium, sulfur, copper, and zinc, increase with temperature, indicating that while nitrogen is lost, other micronutrients become more concentrated in the biochar [66]. Elements such as potassium and phosphorus tend to increase at pyrolysis temperatures of 600 °C, which is attributed to the low volatility of potassium oxides (K_2O) and phosphorus pentoxide (P_2O_5) during pyrolysis [71].

Physicochemical properties, such as pH, EC, zero charge point, and surface charge, are useful indicators for biochar quality and can be easily applied at an industrial scale [9]. As observed in biochar characteristics, bonechar quality is determined by production conditions, such as temperature and residence time [62].

The yield of bonechar is directly influenced by pyrolysis temperature, as the temperature increases, the yield of bonechar decreases [28]. Li et al. [91] observed that lower temperatures (such as 300 °C) tend to result in higher mass yield since less organic material is converted into gases. In contrast, higher temperatures (such as 700 and 900 °C) can lead to greater mass loss due to the decomposition of organic compounds, resulting in a lower overall yield of bonechar. This decrease in yield is attributed to the progressive decomposition of the volatile fraction of the biomass at higher temperatures, leading to the transformation of part of the biomass into gaseous products and bio-oil. Ash content in bonechar significantly increases with the rise in pyrolysis temperature [91]. Ash content is an important aspect to consider, especially when evaluating the quality of the produced bonechar [61].

The physical properties of bonechar, such as pH and electrical conductivity, increase with pyrolysis temperatures [3, 12, 13, 91]. Typically, electrical conductivity varies with temperature changes, which can influence the performance of the char in adsorption applications [91]. Alkurdi et al. [9] observed that as pyrolysis temperature increased from 500 to 900 °C, the pH of the bonechar also increased, from 8.4 to 11.6. The increase in pH with temperature is attributed to the loss of acidic functional groups and the increase in alkaline aromatic groups [9].

Száková et al. [153] observed that higher temperatures were associated with greater nutrient mobility in bonechar, except for calcium and phosphorus. This suggests that although bonechar contains a higher concentration of nutrients, their availability may be influenced by the temperature at which it is produced. Amin [14] observed that as temperature increases, the total content of calcium, phosphorus, and sodium in bonechar generally increases from 350 to 750 °C. Bonechar treated by pyrolysis at 500 °C showed the most substantial increase in phosphorus availability compared to untreated soil, indicating its potential as a slow-release fertilizer [12, 13]. Overall, while pyrolysis at lower temperatures (350 °C) increases certain forms of extractable phosphorus, higher temperatures (750 °C) tended to decrease extractable phosphorus in water due to increased crystallization, which limits the dissolution of calcium phosphate in the soil [182].

4 Environmental and Economic Relevance of Biochar and Bonechar

The circular economy offers an alternative to the linear model of "extract–produce–discard," which is considered unsustainable. This concept promotes sustainable development, encouraging the responsible and cyclical use of resources. The goal is to preserve the value of materials in the economy and reduce environmental pressures [120]. The circular economy is a production and consumption model that aims to optimize resource use, reduce waste generation, and promote the regeneration of natural systems [31]. To better understand this concept, Fig. 2 presents a diagram illustrating the main material flows and strategies for closing production cycles, highlighting the stages of reuse, recycling, and resource recovery.

By maximizing the use of resources, minimizing waste, and promoting the reuse and recycling of materials, the circular economy helps reduce pressure on natural resources and decreases the amount of waste generated. Recently, with the increase in public awareness, the focus on developing a circular economy model has intensified,

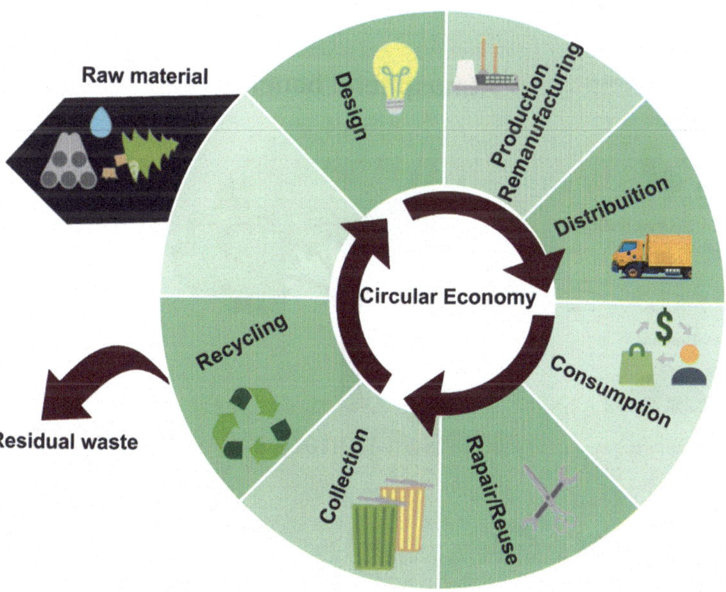

Fig. 2 Informative diagram of a circular economy system for waste recycling. *Source* Authors (2024)

with an emphasis on proper waste management to bring benefits to the agricultural sector [44]. In this context, the objectives of the circular economy paradigm include waste elimination, product and material circularity, and their regeneration [88]. The growing production of by-products is a challenge for modern society, and their valorization by transforming them into valuable compounds is essential for the circular economy. In this context, the integration of chars, such as biochar and bonechar, into agricultural practices and waste management promotes resource reuse, reducing waste [67, 125, 126]. The production of biochar uses biomass waste (Fig. 3), transforming it into a valuable product, and this process aligns with the principles of the circular economy by converting waste into resources, reducing the need for landfills, and promoting sustainable waste management practices [87].

Biochar has gained popularity recently in the environmental sector as a versatile material for waste reduction and increased efficiency in the circular economy, as its applications have demonstrated great potential to reduce environmental impacts, combat climate change, and create an efficient circular economy model [144]. It offers numerous economic and environmental benefits, and its use in soil can significantly improve health and fertility, as well as facilitate bioremediation. Figure 3 summarizes the benefits of biochar both for the soil and the environment.

Biochar is recognized for its ability to remediate soil, water, and air contaminants [176, 177]. Its application to soil brings significant benefits, such as reducing nutrient leaching [119] and improving cation exchange capacity, pH, water retention, as well as decreasing the need for fertilizers and greenhouse gas emissions [16, 17, 43, 180]. Oxygen-containing groups on the biochar surface help in the adsorption

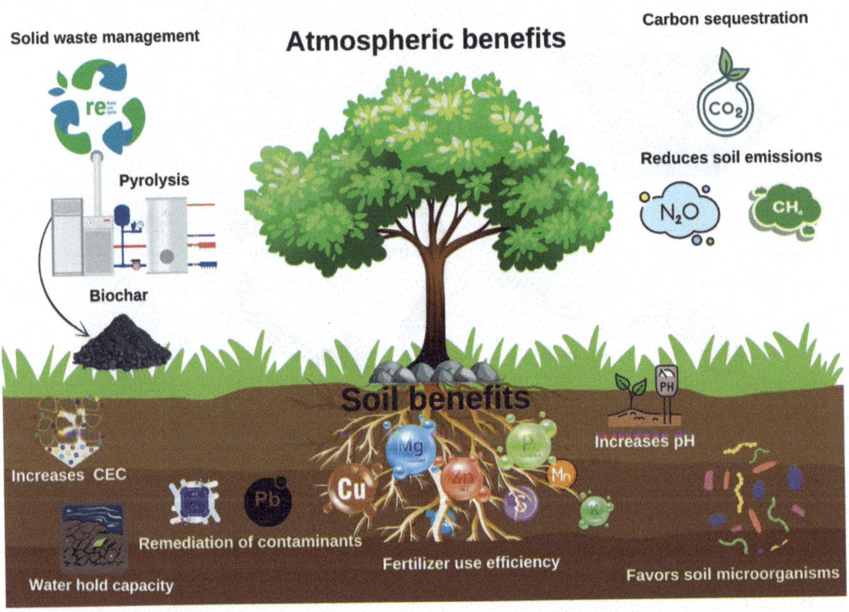

Fig. 3 Benefits of biochar for the atmosphere and soil. *Source* Authors (2024)

of organic pollutants and heavy metals, contributing to soil quality and reducing pollution [33, 150].

In addition to environmental benefits, biochar represents a significant economic opportunity. Harsono et al. [59] reported that in Malaysia, biochar production costs were estimated at $532.00 per year, with an annual revenue of $8,012.00, indicating a profitable return. Nematian et al. [115] showed in California, USA, that biochar production could contribute approximately $1.3 million annually to the local income, impacting the Gross Domestic Product (GDP) between $2.1 million and $5.2 million. The global biochar market is expected to reach $368.65 million by 2028, reflecting the growing recognition of its benefits [88].

On the other hand, bonechar transforms discarded bone waste into a valuable resource [9, 60]. Bonechar is useful for water treatment and promotes the circular economy, reducing environmental impact [9, 60, 61]. It facilitates phosphorus recycling, reducing dependence on mineral fertilizers and mitigating the environmental impacts of chemical fertilizer use [65]. Thus, both biochar and bonechar are waste conversion technologies that promote sustainability and resource use efficiency. They also have significant applications for agriculture and environmental remediation: bonechar has the ability to immobilize heavy metals, such as zinc and cadmium, reducing their availability to plants and consequently their toxicity in soil [17, 61]. Moreover, bonechar is a renewable source of phosphorus [11, 21, 55, 168], considered one of the most limiting nutrients for plant growth and development, and can serve as an alternative strategy to meet fertilizer demand, particularly phosphorus, in developing countries where traditional phosphorus sources may be limited [168], especially in acidic soils common in the tropics [55].

The use of bonechar can improve soil properties, including pH levels, which, in turn, can lead to improved nutrient availability and overall soil fertility [51]. Long-term application has been associated with increased agricultural productivity, making it a viable option for sustainable agriculture [168]. Another crucial point is that the use of bonechar emerges as a promising solution for removing various pollutants in water treatment management, with heavy metals such as nickel, cadmium, chromium, lead, copper, arsenic, and fluoride being cited as harmful to health when humans are exposed to concentrations above recommended limits [51]. This capability positions it as a low-cost and efficient solution to address water quality issues, especially in developing regions where access to clean drinking water is a critical concern [100, 113]. Table 1 summarizes the main characteristics and functionalities of biochar and bonechar.

The economic importance of both materials goes beyond their direct applications in agriculture and environmental remediation. Biochar and bonechar are produced from waste, such as agricultural and animal residues, creating an opportunity for the valorization of these wastes. The use of bonechar as an adsorbent material emerges as an efficient and safe alternative for the treatment of these wastes, and in addition to being an economical practice, bones are widely available [121]. The thermal conversion of fish bones, which are often discarded as waste, can generate new sources of income for fishing communities, increasing their economic resilience and ability to adapt to environmental challenges such as climate change [27].

Table 1 Main applications of biochar and bonechar in environmental remediation, with an emphasis on the sorption of herbicides, heavy metals, and other contaminants, as well as their contribution to improving soil and water quality

Material	Application	Main function	Examples of herbicides or pollutants	Reference
Biochar	Herbicide leaching in soil	It acts as a sorption agent that reduces the leaching of herbicide	Imazapic, imazapyr, and onduty (a mixture of herbicides containing imazapic and imazapyr)	Yavari et al. (2021)
Bonechar	Herbicide immobilization in soil	Bonechar acts as a sorption agent that decreases the availability of the herbicide in the soil	Indaziflam	Mendes et al. [105]
Bonechar	Removal of pesticides from contaminated drinking water	It acts as an adsorbent for the removal of pesticides from drinking water	Hexazinona Diurona Ametryn Sulfometuron-methyl Azoxystrobin	Mendes et al. [106]
Bonechar	Sorption–desorption and mobility of herbicides in soil	It acts as an adsorbent to increase sorption and reduce the mobility of herbicides in the soil	Hexazinone, metribuzin, and quinclorac	Mendes et al. [104]
Biochar	Retention and mobility of herbicides, soil quality improvement, and mitigation Heavy metal contamination	Reduce nutrient leaching and mitigate contamination through the sorption of heavy metals or organic pollutants	Herbicides (Monuron, Diuron, and Linuron) Heavy metals in soil (lead, cadmium, arsenic)	Wang et al. [172]

(continued)

Table 1 (continued)

Material	Application	Main function	Examples of herbicides or pollutants	Reference
Biochar	Soil amendment to improve herbicide absorption	Improve herbicide sorption to reduce its mobility and minimize the risk of contamination of surface and groundwater	Herbicides (pretilachlor and pendimethalin)	Thilagavathi et al. [157]
Biochar	Reduce the absorption of herbicide residues	Mitigate the effects of herbicide residues by adsorbing them in the soil, reducing their availability and toxicity to plants	Glyphosate	Singha et al. [146]
Bonechar	Removal of herbicide from contaminated water	Its high sorption capacity allows for the retention of herbicides and other pollutants in aqueous solutions	Metribuzin	Meilke et al. [108]
Bonechar	Improve phosphorus availability in acidic and alkaline soils, as well as promote the remediation of soils contaminated by heavy metals	Acts as a source of phosphorus, improves soil quality, remediates contaminated soils, and reduces nutrient leaching	Heavy metals (lead, cadmium, copper, and zinc)	Amin [15]

(continued)

Table 1 (continued)

Material	Application	Main function	Examples of herbicides or pollutants	Reference
Bonechar	Adsorbent for the removal of contaminants from water	It acts as an adsorbent for fluoride removal from water, due to the presence of hydroxyapatite in its composition, which enables ion exchanges and efficient adsorption of contaminants	Fluoride (F^-) Cadmium and triclosan	Alkurdi et al. [7], Fung et al. [54], González-Ponce et al. [56], Cisneros-Ontiveros et al. [30]
Bonechar	Removal of inorganic arsenic species	It acts as an adsorbent to remove inorganic arsenic species (As(III) and As(V)) from water, interacting with the ions present in the aqueous solution	Arsenic (As(III) and As(V))	Alkurdi et al. [8]

Source The Authors (2024)

5 Use of Biochar and Bonechar in Potable Water and Wastewater Treatment

The increasing demand for high-quality water resources and the rise in industrial, agricultural, and domestic effluent production have heightened the need for effective and affordable water treatment technologies. Among the various approaches explored, the use of bio-based adsorbents, such as biochar and bonechar, stands out due to their ability to sustainably remove organic and inorganic contaminants, aligning with the principles of the circular economy. As previously discussed, both materials possess high surface area, significant adsorption capacity, and porosity, which also facilitate the capture of metal ions present in water.

Studies have shown that heavy metals such as lead, cadmium, zinc, mercury, arsenic, and copper are efficiently removed from effluents through interactions with the oxygenated functional groups on the surfaces of biochar and bonechar [74, 112, 181]. These interactions occur via mechanisms of chemisorption and complexation, where metals form complexes with carboxyl, hydroxyl, and carbonyl groups of the biochar or bonechar [18]. This not only prevents water contamination by these toxic metals but also offers a viable and low-cost alternative for treating industrial and domestic effluents, particularly in regions with limited infrastructure.

The use of biochar and bonechar in water treatment systems is widely integrated into filter-based technologies, leveraging their adsorptive properties to enhance the quality of treated water. These filters have been the focus of numerous studies aimed at optimizing their efficiency and applicability in various contexts, ranging from simple household systems to complex industrial operations.

Studies have explored the use of multilayer filters, in which biochar or bonechar is combined with materials such as sand or zeolite to enhance adsorption and filtration capacity. These filters are effective in removing organic and inorganic contaminants, as well as pathogens, resulting in high-quality treated water [86, 130]. In rural communities or resource-limited areas, slow-flow filters with bonechar have proven to be practical and low-cost solutions. These systems rely on gravity to allow prolonged contact between contaminated water and bonechar, facilitating the adsorption of substances that, in excess, are considered harmful to humans, such as nitrates, phosphates, and fluoride [54]. Additionally, chemical or physical modifications of biochar and bonechar, such as iron impregnation or pH adjustment, have significantly enhanced the removal of specific contaminants. For instance, iron-impregnated biochar filters are effective in removing arsenic due to their ability to form stable complexes with the contaminant [4, 63].

The benefits of these filters include sustainability, as they utilize organic and animal waste, promoting sustainable waste management; cost efficiency, being more affordable and requiring less maintenance compared to conventional technologies; and application flexibility, as they can be adapted for various scales, from household systems to industrial plants.

5.1 Use of Biochar and Bonechar in Wetland Systems

The United Nations (UN) Sustainable Development Goal 6—"Clean Water and Sanitation"—aims to ensure the availability and sustainable management of water, as well as sanitation for all, by 2030 [173]. In this context, many wetlands (wetland systems—WSs) have been constructed and studied over recent decades, primarily for wastewater and drinking water treatment [38], stormwater runoff management, or to mitigate the loss of natural wetlands. WSs are engineered natural systems that combine vegetation (emergent, floating, or submerged), substrates (rocks, soils, synthetic materials, among others), and microbial communities in controlled environments, where pollutants from effluents are removed through the combined effect of physical, chemical, and biological processes [167]. WSs have gained popularity in recent years due to their relatively simple and cost-effective construction, operation, and maintenance [48].

Studies have shown that the traditional substrates used as filtering media in constructed wetlands (soil, gravel, sand) primarily serve as plant support, but their effect on contaminant removal is often unsatisfactory [175]. Using these non-reactive

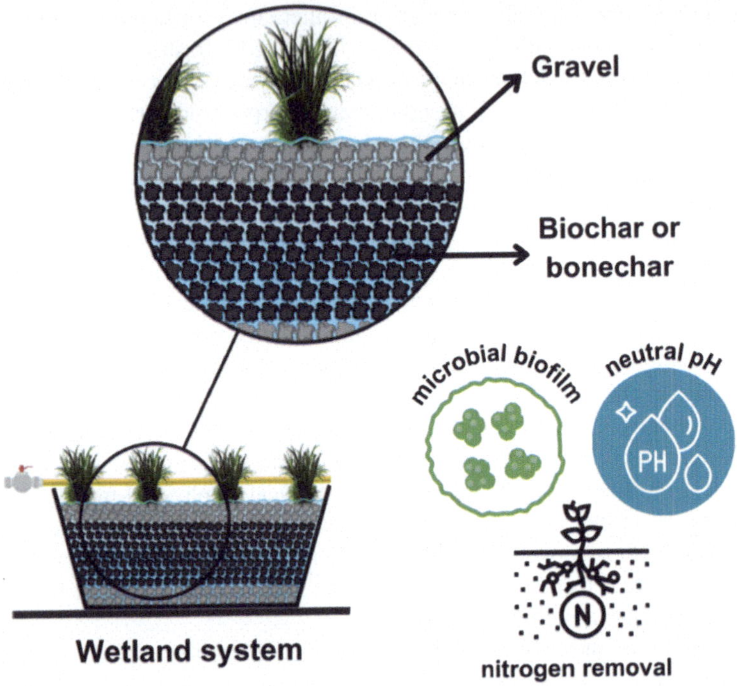

Fig. 4 Wetland systems commonly studied in the literature with biochar or bonechar as filter media and their respective benefits in wastewater treatment. *Source* Authors (2024)

materials to treat high-strength wastewater, such as sewage and industrial effluents, typically requires a relatively large area to achieve sufficient degradation efficiency [84]. With this in mind, biochar and bonechar can be used as substrates in these systems to improve pollutant removal capacity, including organic matter, nutrients (such as nitrogen and phosphorus), heavy metals, and emerging contaminants.

The introduction of biochar and bonechar in wetlands increases the available surface area for contaminant sorption and provides a favorable habitat for the colonization of microbial biofilms (Fig. 4), which assist in the degradation of organic matter. These materials offer a sustainable and affordable alternative for wastewater treatment in rural and peri-urban areas. Another important aspect is that these two adsorbent materials, when saturated and losing their adsorption capacity, can be used as soil amendments due to their nutrient richness, thereby minimizing the disposal issues of residual WS materials in the environment [142].

5.2 Cellulose Nanofilm as an Ally in the Use of Biochar and Bonechar for Herbicide Removal in Water

The use of pure biochar and bonechar in their granular and/or powdered form makes it difficult to remove them from water, requiring a centrifugation step to remove the matrix after the herbicide has been removed. Additionally, the direct application of carbonaceous materials in water can increase the dissolved carbon content, affecting aquatic ecosystems by raising turbidity levels and the toxicity of metals [94]. One alternative is the use of a composite material, in which fine particles of bonechar and biochar can be added to solid particles through a biopolymer, which may facilitate the collection and/or reuse of the material after the herbicide is removed from the water [108].

Cellulose, the most abundant biopolymer in nature, can be processed at the nanoscale to form nanopellets with high mechanical strength and biodegradability. The incorporation of biochar or bonechar into cellulose nanofilm composites has proven to be a promising alternative to enhance the efficiency of water treatment systems. Mielke et al. [108], working with bonechar produced from bovine bone material incorporated into cellulose acetate film, concluded that the membrane showed the highest sorption potential for the lowest concentrations of the herbicide metribuzin in drinking water. Bonechar added at concentrations of 2 and 3 g adsorbed 40% and 60%, respectively, of metribuzin at concentrations of 0.25, 0.33, and 0.5 mg L^{-1}. For both films, desorption was low, at 7% and 2.5% at 24 and 120 h, respectively. The film becomes an efficient and easy-to-handle alternative for the decontamination of polluted aquatic environments, and can serve as an alternative for water and sewage treatment plants, as well as for use in spray tanks, washing personal protective equipment (PPE), and household filters.

Therefore, these nanopellets act as supports or matrices, improving the structural stability of biochar and bonechar while enhancing pollutant removal efficiency. The interaction between the porous surface of biochar and the cellulose nanofilm favors the retention of organic and inorganic compounds, including pesticides, pharmaceuticals, and other emerging contaminants, due to physical and chemical sorption. The use of this composite can be particularly advantageous in water filtration systems, allowing material regeneration and extending its lifespan in wastewater treatment.

5.3 Biochar Fortification with Iron: An Alternative for Removing Contaminants from Water

Separation materials have received increasing attention due to their wide applications in environmental pollution management. Balancing the contradiction between high separation efficiency and selectivity of separation materials is a key objective. The integration of bonechar, for example, with electrospun membranes can achieve this balance [39]. The modification of biochar with iron has been explored as an efficient strategy for the removal of specific contaminants, such as arsenic and chromium, which require oxidation/reduction processes. When incorporated into the porous surfaces of biochar, iron acts as a catalyst in adsorption reactions, promoting either the oxidation of contaminants or their complexation with iron.

One of the main mechanisms for removing arsenic and chromium with iron-modified biochars involves the adsorption of these metals onto active surfaces, where iron facilitates both the oxidation of chromium from chromium (VI) to chromium (III), a less toxic form, and the adsorption of arsenic species [85, 128, 148]. Generally, iron-modified biochars result in a significant increase in the adsorption capacity for arsenic and chromium (VI) compared to unmodified biochars. These improvements are attributed to the interactions between the oxygenated functional groups of biochar and the impregnated iron species, which help create more effective adsorption sites for capturing these contaminants.

In a study by Nair et al. [114], the application of ammonia-environment pyrolysis enhanced the magnetism of biochar. This enhancement facilitated the efficient separation and collection of biochar from the solution after chromium (VI) treatment. The magnetic property introduced through this process provides a practical advantage during the post-treatment phase, streamlining the removal of biochar from the treated solution. These modifications ensure efficient removal while enabling the regeneration and reuse of biochar, making the process more sustainable. These advances are part of a global strategy to develop more effective and affordable materials for the remediation of contaminated water, particularly in regions with high levels of heavy metal pollution, such as Asia and Europe.

6 Addition of Biochar and Bonechar to Immobilize Residual Herbicides in Soil and Water

Environmental protection and sustainable land use practices have increasingly gained prominence in global debates. International sustainability conferences promote new ideas for preserving nature. However, population growth increases the demand for food, intensifying land use and the application of herbicides [98]. Excessive application of these agrochemicals can contaminate the soil, resulting in damage that is difficult or even impossible to reverse [102]. Herbicides like atrazine, in high concentrations, impair plant growth [64], and the growing use of such substances poses serious environmental risks.

This concern is heightened by the leaching potential of pesticides, which can contaminate groundwater and other water resources [40]. According to Santos et al. [137], herbicides such as glyphosate, paraquat dichloride, and residual atrazine exhibit high contamination potential, both in dissolved water forms and when associated with sediments. The solubility of these compounds, often expressed in parts per million (ppm), directly influences their leaching and contamination behavior [93].

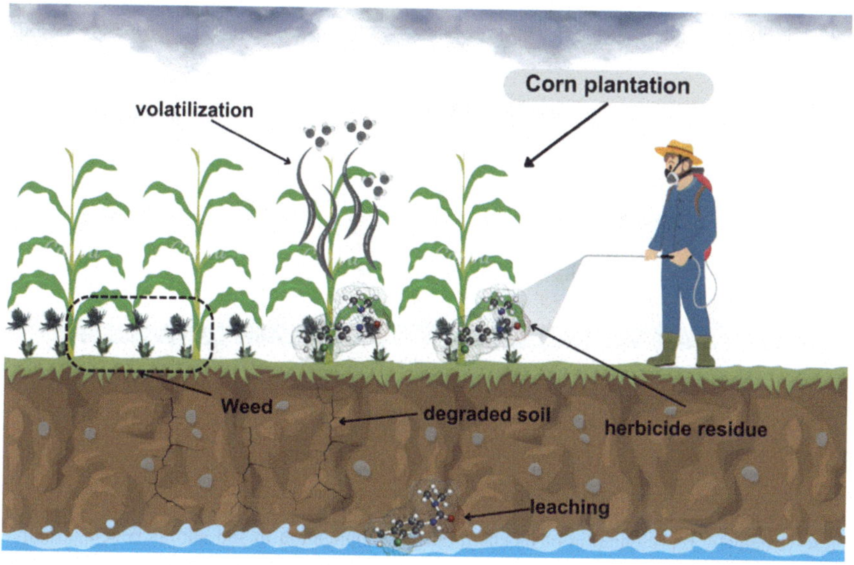

Fig. 5 Schematic representation of environmental contamination due to the use of highly leachable pesticides. *Source* Authors (2024)

Excessive herbicide application affects soil, water, and air, making it essential to understand the environmental risks they pose (Fig. 5). Many countries, especially in Europe, have intensified their focus on pesticide use and its impact on groundwater, which constitutes a significant portion of drinking water supplies. Studying and discussing the dynamics of this process is crucial to mitigating its effects [47].

Biochar, in turn, is a promising alternative for environmental remediation, being used in soil for carbon storage and water filtration. In addition to being economically viable and durable, it has the capacity to adsorb organic and inorganic pollutants, such as herbicides [119]. For instance, biochars produced from coffee processing residues demonstrate high sorption capacity for herbicides like atrazine [32]. This practice also improves soil quality, enhances agricultural productivity, and contributes to waste management and climate change mitigation (Fig. 6) [90].

Moreover, bonechar, produced from bones, offers a rich source of phosphorus, making it useful in fertilizers and various applications such as environmental remediation and sustainable agriculture [117, 126]. Its use is also effective in removing pollutants, including heavy metals and organic substances.

Biochar, depending on the biomass and pyrolysis conditions, can influence the sorption and degradation of herbicides, such as metribuzin, proving effective in the recovery of contaminated soils [110]. Its ability to reduce the mobility of organic pollutants helps mitigate environmental damage caused by herbicides [147]. In a study conducted by Ferreira Mendes et al. [106], bovine bone char was able to adsorb pesticides such as hexazinone, diuron, ametryn, sulfometuron-methyl, and azoxystrobin from contaminated drinking water. The method used in this study was

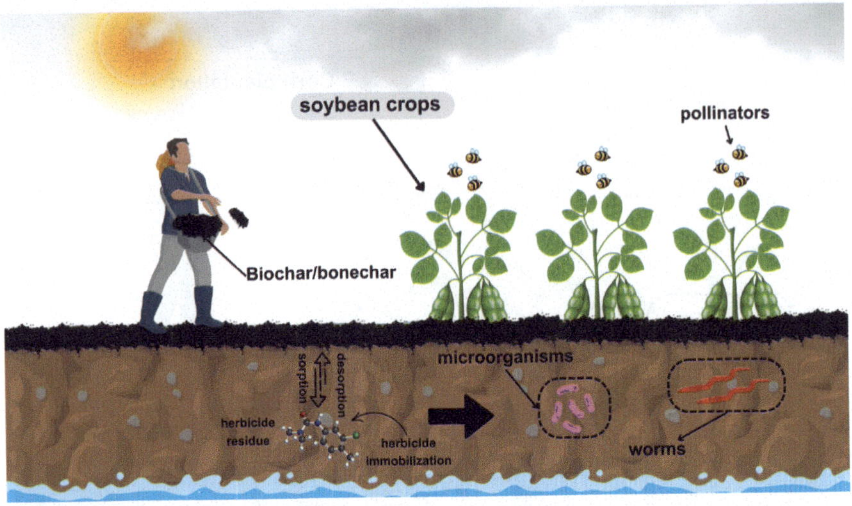

Fig. 6 Schematic representation of biochar/bonechar application in soil and its benefits. *Source* Authors (2024)

practical, indicating that the use of bone char for removing pesticides from drinking water could be scaled up for use in water treatment plants or domestic filters.

Both biochar and bonechar are effective in removing herbicides from soils and water bodies contaminated by leaching and surface runoff. These materials adsorb pesticide molecules within their pores, and their ability to regenerate after adsorption makes them a practical and sustainable solution for treating agricultural effluents.

This efficient use of natural resources reduces reliance on synthetic chemical inputs, promoting more resilient and eco-friendly agricultural practices. Sustainable agriculture supported by the use of biochar and bonechar not only enhances food security by increasing crop productivity and quality but also helps build agricultural systems more resilient to climate change. By improving soil health and sequestering carbon, these practices contribute to climate change mitigation, creating a virtuous cycle that benefits the environment and food production in the long term.

7 Public Health: Issues and Potential Solutions Involving the Use of Biochar and Bonechar

Although agricultural commodities are essential to the global economy, the benefits of pesticide use in crop preservation must be carefully weighed against the risks they pose to human health and the environment. Therefore, it is crucial to promote scientific studies that evaluate diseases associated with pesticide use [123].

The effectiveness of pesticides in inhibiting harmful pests is undeniable. However, their use has caused serious harm to the environment and human health, contaminating soil and water bodies and impacting fauna and flora directly or indirectly. Moreover, exposure to agrochemicals is linked to various diseases, including cancer, diabetes mellitus, and reproductive, respiratory, and neurological disorders [131]. Considering that glyphosate is one of the most widely used herbicides in Brazil and globally, and given the concerns about its toxicity, the number of publications on this topic grows yearly [118]. Reports associating glyphosate with cancer risks, endocrine disruption, celiac disease, autism, effects on erythrocytes, leaky gut syndrome, and other disorders highlight the need to consider its impacts [101].

During periods of heavy rain, pesticides infiltrate the soil, reaching aquifers and riverbeds, compromising fauna and ecosystems [143]. Herbicides can cause several agricultural production issues, including increased production costs due to the need for more qualified professionals, risks of human intoxication, contamination of aquifers and ecosystems, and the emergence of resistant biotypes [35]. The excessive use of agrochemicals results in frequent incidents, including direct and occupational contamination and ingestion of food containing residues. The environment also suffers significant damage. In Brazil, for example, cases of poisoning in breastfeeding mothers and pesticide spraying on children and adolescents in schools highlight critical factors requiring further study of the effects and solutions to mitigate problems arising from pesticide use both in Brazil and worldwide [41].

The improper disposal of organic waste, such as bones and sugarcane bagasse, can lead to severe environmental and public health problems, including soil and groundwater contamination, proliferation of disease vectors, and greenhouse gas emissions. However, transforming this waste into biochar and bonechar offers sustainable solutions, such as nutrient recycling, emission mitigation, soil quality improvement, and promotion of public health. These practices align with the circular economy, fostering agricultural and environmental sustainability and contributing to developing technologies that benefit rural and urban communities.

8 Challenges and Future Perspectives

Although biochar and bonechar hold significant potential for water treatment and herbicide immobilization in soil, challenges still need to be addressed, particularly concerning the standardization of production processes and the efficiency of the adsorbents under different environmental conditions. The variability in biochar properties, which depends on the biomass used and the pyrolysis conditions, can significantly impact its performance. Additionally, the regeneration of adsorbents after the saturation of pollutants is a crucial issue for the feasibility of large-scale application.

From a social perspective, the adoption of these technologies can contribute to food security and the improvement of living conditions in rural communities by promoting more sustainable agricultural practices. The use of biochar and bonechar can help increase agricultural productivity and soil quality, benefiting small producers and vulnerable communities who depend on farming for their livelihood. Furthermore, these practices are aligned with the circular economy, as proposed by the Brazilian Bill 1874/2022, which encourages the valorization of agro-industrial and forestry residues, promoting more sustainable resource management.

Decarbonization is also an important consideration, as the use of biochar and bonechar can help reduce greenhouse gas emissions through carbon sequestration in the soil. To ensure the effectiveness and safety of these solutions, it is essential for future research to focus on optimizing production conditions, following quality standards such as the European Biochar Certificate [139], which establishes guidelines for sustainable biochar production.

Moreover, the evaluation of the efficiency of these materials in real water and soil treatment systems, as well as the integration of these technologies with complementary approaches such as constructed wetlands and disinfection processes, will be essential for the development of sustainable and cost-effective treatment solutions. This integrated approach can not only address environmental challenges but also generate positive social impacts, promoting inclusion and resilience in affected communities.

9 Conclusion

With the increasing production demand of agro-industries, the intensive use of herbicides and chemical inputs has caused serious environmental impacts, such as soil and water contamination. These pollutants not only harm ecosystem health but also compromise agricultural sustainability, affecting crop rotation and the health of plants, soil, and water resources. The presence of toxic residues in soil and water has direct implications for biodiversity, quality, and food security. Furthermore, herbicide pollution is a significant contributor to climate change, as highlighted in the IPCC (Intergovernmental Panel on Climate Change) report, which points to the growing pressure of intensive agriculture on ecosystems and the impact of chemical product use on greenhouse gas emissions. Alongside this factor, the intense generation of agro-industrial waste and its improper management also exacerbates this environmental decline, amplifying the damage to the environment and public health. In response to these environmental challenges, the growing demand for sustainable solutions has driven research and the development of innovative technologies such as biochar and bonechar. These materials stand out not only for their ability to remediate contaminated soils and waters but also for their contribution to pollution mitigation and carbon capture, fundamental aspects for the UN's Sustainable Development Goals (SDGs). SDG 6, which aims to ensure the availability and sustainable management of water, and SDG 12, focused on responsible consumption and production, are particularly relevant when considering the use of biochar and bonechar as sustainable alternatives for reusing agro-industrial waste. Moreover, the adoption of these materials helps reduce pressure on natural resources and promotes a circular economy while providing an efficient alternative to replace non-renewable raw materials.

By adopting sustainable practices in agro-industrial waste management, it is possible to minimize environmental impacts, improve soil and water quality, and promote the efficient use of natural resources, aligning with global SDGs. This approach not only contributes to reducing waste and operational costs but also opens new economic opportunities and promotes environmental preservation, boosting ecosystem resilience in the face of climate change and ensuring the well-being of future generations.

References

1. Abdelmagid AAA, Qiu Y (2024) Repurposing agricultural waste: the effectiveness of peanut husk ash in improving asphalt mixture properties. Constr Build Mater 411. https://doi.org/10.1016/j.conbuildmat.2023.134476
2. Ahmad M, Lee SS, Dou X, Mohan D, Sung JK, Yang JE, Ok YS (2012) Effects of pyrolysis temperature on soybean stover- and peanut shell-derived biochar properties and TCE adsorption in water. Bioresour Technol 118:536–544. https://doi.org/10.1016/j.biortech.2012.05.042
3. Ahmed M, Nigussie A, Addisu S, Belay B, Lehmann J, Sato S (2023) Valorization of animal bone waste for agricultural use through biomass co-pyrolysis and bio-augmentation. Biomass Convers Biorefin 13(14):12823–12832. https://doi.org/10.1007/s13399-021-02100-w
4. Alchouron J, Navarathna C, Rodrigo PM, Snyder A, Chludil HD, Vega AS, Bosi G, Perez F, Mohan D, Pittman CU, Mlsna TE (2021) Household arsenic contaminated water treatment employing iron oxide/bamboo biochar composite: an approach to technology transfer. J Colloid Interface Sci 587:767–779. https://doi.org/10.1016/J.JCIS.2020.11.036
5. Ali I, Asim M, Khan TA (2012) Low cost adsorbents for the removal of organic pollutants from wastewater. J Environ Manage 113:170–183
6. Alkurdi SSA, Al-Juboori RA, Bundschuh J, Hamawand I (2019) Bone char as a green sorbent for removing health threatening fluoride from drinking water. Environ Int 127:704–719. https://doi.org/10.1016/J.ENVINT.2019.03.065
7. Alkurdi SSA, Herath I, Bundschuh J, Al-Juboori RA, Vithanage M, Mohan D (2019) Biochar versus bone char for a sustainable inorganic arsenic mitigation in water: what needs to be done in future research? Environ Int 127:52–69. https://doi.org/10.1016/J.ENVINT.2019.03.012
8. Alkurdi SSA, Al-Juboori RA, Bundschuh J, Bowtell L, Marchuk A (2021) Inorganic arsenic species removal from water using bone char: a detailed study on adsorption kinetic and isotherm models using error functions analysis. J Hazard Mater 405:124112. https://doi.org/10.1016/j.jhazmat.2020.124112
9. Alkurdi SSA, Al-Juboori RA, Bundschuh J, Bowtell L, McKnight S (2020) Effect of pyrolysis conditions on bone char characterization and its ability for arsenic and fluoride removal. Environ Pollut 262. https://doi.org/10.1016/j.envpol.2020.114221
10. Amalina F, Razak ASA, Krishnan S, Sulaiman H, Zularisam AW, Nasrullah M (2022) Biochar production techniques utilizing biomass waste-derived materials and environmental applications—a review. J Hazard Mater Adv 7. https://doi.org/10.1016/j.hazadv.2022.100134
11. Amin AEEAZ (2021) Enhancement of releasing phosphorus from bone char in calcareous sandy soil under applying different levels of water salinity. J Soil Sci Plant Nutr 21(1):476–486. https://doi.org/10.1007/S42729-020-00376-X/FIGURES/4
12. Amin AEEAZ (2023) Effects of pyrolysis temperatures on bone char characterization and its releasing phosphorus in sandy soil. Arch Agron Soil Sci. https://doi.org/10.1080/03650340.2021.1988940

13. Amin AEEAZ (2023) Effect of co-applying different nitrogen fertilizers with bone char on enhancing phosphorus release in calcium carbonate-rich soil: an incubation study. J Soil Sci Plant Nutr 23(2):1565–1575. https://doi.org/10.1007/s42729-023-01217-3
14. Amin AEEAZ (2023) Effects of pyrolysis temperatures on bone char characterization and its releasing phosphorus in sandy soil. Arch Agron Soil Sci. https://doi.org/10.1080/03650340.2021.1988940 [MM1]
15. Amin AE-EAZ (2024) Using bone char as a renewable resource of phosphate fertilizers in sustainable agriculture and its effects on phosphorus transformations and remediation of contaminated soils as well as the growth of plants. J Soil Sci Plant Nutr. https://doi.org/10.1007/s42729-024-02018
16. Antoniou N, Monlau F, Sambusiti C, Ficara E, Barakat A, Zabaniotou A (2019) Contribution to circular economy options of mixed agricultural wastes management: coupling anaerobic digestion with gasification for enhanced energy and material recovery. J Clean Prod 209:505–514. https://doi.org/10.1016/j.jclepro.2018.10.055
17. Azeem M, Ali A, Arockiam Jeyasundar PGS, Bashir S, Hussain Q, Wahid F, Ali EF, Abdelrahman H, Li R, Antoniadis V, Rinklebe J, Shaheen SM, Li G, Zhang Z (2021) Effects of sheep bone biochar on soil quality, maize growth, and fractionation and phytoavailability of Cd and Zn in a mining-contaminated soil. Chemosphere 282. https://doi.org/10.1016/j.chemosphere.2021.131016
18. Azeem M, Shaheen SM, Ali A, Jeyasundar PGSA, Latif A, Abdelrahman H, Li R, Almazroui M, Niazi NK, Sarmah AK, Li G, Rinklebe J, Zhu YG, Zhang Z (2022) Removal of potentially toxic elements from contaminated soil and water using bone char compared to plant- and bone-derived biochars: a review. J Hazard Mater 427. https://doi.org/10.1016/j.jhazmat.2021.128131
19. Bansal RK, Gondaliya VK, Shaikh AS (2017) A review of the status of the groundnut production and export of India. Indian J Econ Dev 13(2):369. https://doi.org/10.5958/2322-0430.2017.00190.1
20. Bartoli M, Giorcelli M, Jagdale P, Rovere M, Tagliaferro A (2020) A review of non-soil biochar applications. Materials 13
21. Bayata A, Mulatu G (2024) Effect of bone char application on soil quality, soil enzyme and in enhancing crop yield in agriculture: a review. Am J Chem Eng 12(2):13–28. https://doi.org/10.11648/j.ajche.20241202.11
22. Bhattacharyya P, Bisen J, Bhaduri D, Priyadarsini S, Munda S, Chakraborti M, Adak T, Panneerselvam P, Mukherjee AK, Swain SL, Dash PK, Padhy SR, Nayak AK, Pathak H, Arora S, Nimbrayan P (2021) Turn the wheel from waste to wealth: economic and environmental gain of sustainable rice straw management practices over field burning in reference to India. Sci Total Environ 775
23. CONAB (2022) Acompanhamento da safra brasileira de cana-de-açúcar, 2022/2023, 2º levantamento, vol 9, pp 1–58
24. CONAB (2023) Produção de amendoim cresce mais de 100% nos últimos 8 anos. https://www.conab.gov.br/ultimas-noticias/4768-producao-de-amenm-cresce-mais-de-100-nos-ultimos-8-anos. Acesso em: 29 de Julho de 2024
25. CONAB—COMPANHIA NACIONAL DE ABASTECIMENTO. Acompanhamento de safra brasileira de grãos. Brasília, DF
26. Cantrell KB, Hunt PG, Uchimiya M, Novak JM, Ro KS (2012) Impact of pyrolysis temperature and manure source on physicochemical characteristics of biochar. Bioresour Technol 107:419–428. https://doi.org/10.1016/j.biortech.2011.11.084
27. Carella F, Seck M, Esposti LD, Diadiou H, Maienza A, Baronti S, Vignaroli P, Vaccari FP, Iafisco M, Adamiano A (2021) Thermal conversion of fish bones into fertilizers and biostimulants for plant growth-A low tech valorization process for the development of circular economy in least developed countries. J Environ Chem Eng 9(1). https://doi.org/10.1016/j.jece.2020.104815
28. Chagtmi R, Trabelsi ABH, Haddad K, Maaoui A, Lopez G, Santamaria L, Cortazar M, Khedira H, Chaden C, Olazar M (2022) Biochar derived from animal bone waste as an alternative

bioadsorbent for an industrial tannery dye removal. Biomass Convers Biorefin. https://doi.org/10.1007/s13399-022-03459-0
29. Chan KY, Van Zwieten L, Meszaros I, Downie A, Joseph S (2007) Agronomic values of greenwaste biochar as a soil amendment. Aust J Soil Res 45(8):629–634. https://doi.org/10.1071/SR07109
30. Cisneros-Ontiveros HG, Medellín-Castillo NA, Flores-Rojas AI, Cruz-Briano SA (2024) Removal of fluoride, cadmium and triclosan from water using bone chars from an invasive species: optimization, equilibrium and adsorption mechanisms. Sustain Chem Environ 8:100182. https://doi.org/10.1016/j.scenv.2024.100182
31. Corona B, Shen L, Reike D, Rosales Carreón J, Worrell E (2019) Towards sustainable development through the circular economy—a review and critical assessment on current circularity metrics. Resour Conserv Recycl 151:104498. https://doi.org/10.1016/J.RESCONREC.2019.104498
32. da Silva LFV, Pinto AA, Mendonça JCF, do Prado JR, de Melo EI, de Paula GH, dos Junior PJS (2020) Efeito do biochar de resíduo do fruto do cafeeiro na sorção de atrazina no solo. Magistra 31:543–553
33. Das S, Mohanty S, Sahu G, Rana M, Pilli K (2021) Biochar: a sustainable approach for improving soil health and environment. In: Vieira A, Rodrigues SC (eds) Soil erosion. IntechOpen, Rijeka
34. de Benedicto DFC, Ferreira GMD, Thomasi SS (2024) β-cyclodextrin-functionalized coffee husk biochar for surfactant adsorption. Colloids Surf A Physicochem Eng Asp 701. https://doi.org/10.1016/j.colsurfa.2024.134921
35. de Medeiros EF (2023) Diagnóstico de Utilização de Herbicidas no Município de Ceará-Mirim, RN. Trabalho de Conclusão de Curso (TCC), Universidade Federal do Rio Grande do Norte
36. Deng S, Niu L, Bei Y, Wang B, Huang J, Yu G (2013) Adsorption of perfluorinated compounds on aminated rice husk prepared by atom transfer radical polymerization. Chemosphere 91(2):124–130. https://doi.org/10.1016/j.chemosphere.2012.11.015
37. Domingues RR, Trugilho PF, Silva CA, De Melo ICNA, Melo LCA, Magriotis ZM, Sánchez-Monedero MA (2017) Properties of biochar derived from wood and high-nutrient biomasses with the aim of agronomic and environmental benefits. PLoS One 12(5). https://doi.org/10.1371/journal.pone.0176884
38. Donde OO, Xiao B (2017) Understanding wastewater treatment mechanisms: a review on detection, removal, and purification efficiencies of faecal bacteria indicators across constructed wetlands, vol 25, no 4, pp 444–451. https://doi.org/10.1139/ER-2017-0017
39. Du X, Rashid SA, Abdullah LC, Rahman NA (2023) Fabrication of electrospun cellulose/chitosan/ball-milled bone char membranes for efficient and selective sorption of Pb(II) from aqueous solutions. Environ Sci Pollut Res 30(51):110417–110430. https://doi.org/10.1007/s11356-023-30213-4
40. Dugan ST, Muhammetoglu A, Uslu A (2023) A combined approach for the estimation of groundwater leaching potential and environmental impacts of pesticides for agricultural lands. Sci Total Environ 901:165892. https://doi.org/10.1016/J.SCITOTENV.2023.165892
41. Dutra RMS, Souza MMO (2017) Impactos negativos do uso de agrotóxicos à saúde humana. Revista Brasileira de Geografia Médica e da Saúde-Hygeia 13(24):127–140
42. Echeverria MC, Nuti M (2017) Valorisation of the residues of coffee agro-industry: perspectives and limitations. Open Waste Manage J 10(1):13–22. https://doi.org/10.2174/1876400201710010013
43. Enaime G, Lübken M (2021) Agricultural waste-based biochar for agronomic applications. Appl Sci 11(19):8914. https://doi.org/10.3390/APP11198914
44. Enaime G, Wichern M, Lübken M (2023) Contribution of biochar application to the promotion of circular economy in agriculture. Front Agron 5. https://doi.org/10.3389/fagro.2023.1214012
45. FAO (2024) FAOSTAT—data. https://www.fao.org/faostat/en/#data/QCL. Acesso em: 1 de julho de 2024

46. FEDERAÇÃO DAS ASSOCIAÇÕES RURAIS DO ESTADO DE SÃO PAULO (FAESP) (2024) Perspectiva para o mercado de açúcar e etanol. FAESP, São Paulo. https://faespsenar.com.br. Acesso em: 01 de junho de 2024
47. Fenoll J, Ruiz E, Flores P, Vela N, Hellín P, Navarro S (2011) Use of farming and agro-industrial wastes as versatile barriers in reducing pesticide leaching through soil columns. J Hazard Mater 187(1–3):206–212. https://doi.org/10.1016/J.JHAZMAT.2011.01.012
48. Fernández del Castillo A, Garibay MV, Senés-Guerrero C, Orozco-Nunnelly DA, de Anda J, Gradilla-Hernández MS (2022) A review of the sustainability of anaerobic reactors combined with constructed wetlands for decentralized wastewater treatment. J Clean Prod 371:133428. https://doi.org/10.1016/J.JCLEPRO.2022.133428
49. Food and Agriculture Organization of the United Nations (2019) The state of food and agriculture 2019. Rome
50. Fornes F, Belda RM (2018) Biochar versus hydrochar as growth media constituents for ornamental plant cultivation. Sci Agric 75(4):304–312. https://doi.org/10.1590/1678-992x-2017-0062
51. Franco De Castro G, Cesar B, Rocha P (2023) Bone char: synthesis, characterization, and agronomic application as an alternative source of phosphorus. https://doi.org/10.21203/rs.3.rs-2828604/v1
52. Freitas LC, Barbosa JR, da Costa ALC, Bezerra FWF, Pinto RHH, de Carvalho Junior RN (2021) From waste to sustainable industry: how can agro-industrial wastes help in the development of new products? Resour Conserv Recycl 169
53. Fu D, Kurniawan TA, Avtar R, Xu P, Othman MHD (2021) Recovering heavy metals from electroplating wastewater and their conversion into Zn2Cr-layered double hydroxide (LDH) for pyrophosphate removal from industrial wastewater. Chemosphere 271. https://doi.org/10.1016/j.chemosphere.2021.129861
54. Fung E, Johnson KI, Li W, Borges W, Chi K, Sharma SK, Madan Y, Sharma PR, Hsiao BS (2021) Study the use of activated carbon and bone char on the performance of gravity sand-bag water filter. Membranes (Basel) 11(11). https://doi.org/10.3390/membranes11110868
55. Glæsner N, Hansen HCB, Hu Y, Bekiaris G, Bruun S (2019) Low crystalline apatite in bone char produced at low temperature ameliorates phosphorus-deficient soils. Chemosphere 223:723–730. https://doi.org/10.1016/J.CHEMOSPHERE.2019.02.048
56. González-Ponce HA, Mendoza-Castillo DI, Bonilla-Petriciolet A, Reynel-Ávila HE, Camacho-Aguilar KI (2023) Regeneration analysis of bone char used in water defluoridation: chemical desorption route, surface chemistry analysis and modeling. Int J Chem Eng 2023. https://doi.org/10.1155/2023/8378162
57. Hadidi M, Aghababaei F, Gonzalez-Serrano DJ, Goksen G, Trif M, McClements DJ, Moreno A (2024) Plant-based proteins from agro-industrial waste and by-products: towards a more circular economy. Int J Biol Macromol 261
58. Hafezi SM, Zarea- Hosseinabadi H, Huijbregts MAJ, Steinmann ZJN (2021) The importance of biogenic carbon storage in the greenhouse gas footprint of medium density fiberboard from poplar wood and bagasse. Cleaner Environ Syst 3. https://doi.org/10.1016/j.cesys.2021.100066
59. Harsono SS, Grundman P, Lau LH, Hansen A, Salleh MAM, Meyer-Aurich A, Idris A, Ghazi TIM (2013) Energy balances, greenhouse gas emissions and economics of biochar production from palm oil empty fruit bunches. Resour Conserv Recycl 77:108–115. https://doi.org/10.1016/j.resconrec.2013.04.005
60. Hart A (2023) Circular economy: closing the catalyst loop with metal reclamation from spent catalysts, industrial waste, waste shells and animal bones. Biomass Convers Biorefin 13(13):11483–11498. https://doi.org/10.1007/s13399-021-01942-8
61. Hart A, Ebiundu K, Peretomode E, Onyeaka H, Nwabor OF, Obileke KC (2022) Value-added materials recovered from waste bone biomass: technologies and applications. RSC Adv 12(34):22302–22330. https://doi.org/10.1039/D2RA03557J
62. Hart A, Porbeni DW, Omonmhenle S, Peretomode E (2023) Waste bone char-derived adsorbents: characteristics, adsorption mechanism and model approach. Environ Technol

Rev 12(1):175–204. https://doi.org/10.1080/21622515.2023.2197128/ASSET/8CB0A6BA-5FF5-456F-9FD1-36487420051A/ASSETS/GRAPHIC/TETR_A_2197128_F0008_OC.JPG
63. He R, Peng Z, Lyu H, Huang H, Nan Q, Tang J (2018) Synthesis and characterization of an iron-impregnated biochar for aqueous arsenic removal. Sci Total Environ 612:1177–1186. https://doi.org/10.1016/J.SCITOTENV.2017.09.016
64. Hernandez MCR (2022) Tolerancia e capacidade remediadora de macrofitas aquáticas ao atrazine. Universidade Federal Rural do Semi-Árido
65. Heyl K, Garske B, Ekardt F (2023) Using bone char as phosphate recycling fertiliser: an analysis of the new EU fertilising products regulation. Environ Sci Eur 35(1). https://doi.org/10.1186/s12302-023-00819-z
66. Hossain MK, Strezov Vladimir V, Chan KY, Ziolkowski A, Nelson PF (2011) Influence of pyrolysis temperature on production and nutrient properties of wastewater sludge biochar. J Environ Manage 92(1):223–228. https://doi.org/10.1016/j.jenvman.2010.09.008
67. Hu Q, Jung J, Chen D, Leong K, Song S, Li F, Mohan BC, Yao Z, Prabhakar AK, Lin XH, Lim EY, Zhang L, Souradeep G, Ok YS, Kua HW, Li SFY, Tan HTW, Dai Y, Tong YW, Peng Y, Joseph S, Wang CH (2021) Biochar industry to circular economy. Sci Total Environ 757. https://doi.org/10.1016/j.scitotenv.2020.143820
68. Huang H, Reddy NG, Huang X, Chen P, Wang P, Zhang Y, Huang Y, Lin P, Garg A (2021) Effects of pyrolysis temperature, feedstock type and compaction on water retention of biochar amended soil. Sci Rep 11(1). https://doi.org/10.1038/s41598-021-86701-5
69. Hussain F, Alshahrani S, Abbas MM, Khan HM, Jamil A, Yaqoob H, Soudagar MEM, Imran M, Ahmad M, Munir M (2021) Review waste animal bones as catalysts for biodiesel production; a mini review. Catalysts 11
70. INSTITUTO BRASILEIRO DE GEOGRAFIA E ESTATÍSTICA (IBGE). Release—Produção pecuária 1° trimestre. https://agenciadenoticias.ibge.gov.br/agencia-sala-de-imprensa/2013-agencia-de-noticias/releases/40001-release-tri-pecuaria-1-tri. Acesso em: 07 out. 2024
71. Ibn Ferjani A, Jeguirim M, Jellali S, Limousy L, Courson C, Akrout H, Thevenin N, Ruidavets L, Muller A, Bennici S (2019) The use of exhausted grape marc to produce biofuels and biofertilizers: effect of pyrolysis temperatures on biochars properties. Renew Sustain Energy Rev 107:425–433. https://doi.org/10.1016/j.rser.2019.03.034
72. Jannat N, Hussien A, Abdullah B, Cotgrave A (2020) Application of agro and non-agro waste materials for unfired earth blocks construction: a review. Constr Build Mater 254
73. Jeguirim M, Bikai J, Elmay Y, Limousy L, Njeugna E (2014) Thermal characterization and pyrolysis kinetics of tropical biomass feedstocks for energy recovery. Energy Sustain Dev 23:188–193. https://doi.org/10.1016/j.esd.2014.09.009
74. Jin Q, Wang Z, Feng Y, Kim YT, Stewart AC, O'Keefe SF, Neilson AP, He Z, Huang H (2020) Grape pomace and its secondary waste management: biochar production for a broad range of lead (Pb) removal from water. Environ Res 186. https://doi.org/10.1016/j.envres.2020.109442
75. Jindo K, Mizumoto H, Sawada Y, Sanchez-Monedero MA, Sonoki T (2014) Physical and chemical characterization of biochars derived from different agricultural residues. Biogeosciences 11(23):6613–6621. https://doi.org/10.5194/bg-11-6613-2014
76. Joseph SD, Camps-Arbestain M, Lin Y, Munroe P, Chia CH, Hook J, Van Zwieten L, Kimber S, Cowie A, Singh BP, Lehmann J, Foidl N, Smernik RJ, Amonette JE (2010) An investigation into the reactions of biochar in soil. Aust J Soil Res 48(6–7):501–515. https://doi.org/10.1071/SR10009
77. Kambo HS, Dutta A (2015) A comparative review of biochar and hydrochar in terms of production, physico-chemical properties and applications. Renew Sustain Energy Rev 45:359–378
78. Kaul K, Rajauria G, Singh R (2024) Food and bioproducts processing valorization of agro-industrial waste for pectinase production and its influence on circular economy. 148(September):141–153

79. Kayranli B, Bilen M, Seckin IY, Yilmaz T, Dinc A, Akkurt F, Simsek H (2024) Peanut shell biochar for Rhodamine B removal: efficiency, desorption, and reusability. Chemosphere 364. https://doi.org/10.1016/j.chemosphere.2024.143056
80. Khater ES, Bahnasawy A, Hamouda R, Sabahy A, Abbas W, Morsy OM (2024) Biochar production under different pyrolysis temperatures with different types of agricultural wastes. Sci Rep 14(1). https://doi.org/10.1038/s41598-024-52336-5
81. Khosravi A, Zheng H, Liu Q, Hashemi M, Tang Y, Xing B (2022) Production and characterization of hydrochars and their application in soil improvement and environmental remediation. Chem Eng J 430:133142. https://doi.org/10.1016/j.cej.2021.133142
82. Kim KH, Kim JY, Cho TS, Choi JW (2012) Influence of pyrolysis temperature on physicochemical properties of biochar obtained from the fast pyrolysis of pitch pine (Pinus rigida). Bioresour Technol 118:158–162. https://doi.org/10.1016/J.BIORTECH.2012.04.094
83. Kim S, Kim JC, Kim YY, Yang JE, Lee HM, Hwang IM, Park HW, Kim HM (2024) Utilization of coffee waste as a sustainable feedstock for high-yield lactic acid production through microbial fermentation. Sci Total Environ 912. https://doi.org/10.1016/j.scitotenv.2023.169521
84. Kizito S, Lv T, Wu S, Ajmal Z, Luo H, Dong R (2017) Treatment of anaerobic digested effluent in biochar-packed vertical flow constructed wetland columns: role of media and tidal operation. Sci Total Environ 592:197–205. https://doi.org/10.1016/j.scitotenv.2017.03.125
85. Kumar P, Singhania RR, Sumathi Y, Kurrey NK, Chen CW, Patel AK, Dong CD (2024) Investigating innovative techniques for biochar modification to enhance the removal of heavy metals from aqueous environments: a comprehensive review. Clean Technol Environ Policy
86. Kumi M (2022) Integrated bone and biochar bed for trace metal, nutrient and pathogen removal from groundwater
87. Kurniawan TA, Liang X, O'callaghan E, Goh H, Othman MHD, Avtar R, Kusworo TD, Kurniawan TA, Liang X, O'callaghan E, Goh H, Hafiz M, Othman D, Avtar R, Kusworo TD (2022) Transformation of solid waste management in China: moving towards sustainability through digitalization-based circular economy. Sustainability 14(4):2374. https://doi.org/10.3390/SU14042374
88. Kurniawan TA, Othman MHD, Liang X, Goh HH, Gikas P, Chong KK, Chew KW (2023) Challenges and opportunities for biochar to promote circular economy and carbon neutrality. J Environ Manage 332. https://doi.org/10.1016/j.jenvman.2023.117429
89. Lee XJ, Ong HC, Gan YY, Chen WH, Mahlia TMI (2020) State of art review on conventional and advanced pyrolysis of macroalgae and microalgae for biochar, bio-oil and bio-syngas production. Energy Convers Manage 210:112707. https://doi.org/10.1016/J.ENCONMAN.2020.112707
90. Lehmann J, Joseph S (2009) Biochar for environmental management: an introduction. In: Earthscan (ed) Biochar for environmental management. Reino Unido, pp 1–12
91. Li Y, Liu J, Han L, Wei B, Liu X (2023) Influence of pyrolysis atmosphere, temperature, and particle size co-regulation on the physicochemical properties of bone char. J Environ Chem Eng 11(6). https://doi.org/10.1016/j.jece.2023.111357
92. Li M, Wang Y, Chen W, Sun Y, Hou H, Liu Y (2024) Assessing GHG emissions of food consumption towards low-carbon transformation in China. Environ Impact Assess Rev 105. https://doi.org/10.1016/j.eiar.2023.107408
93. Lima FB, Reverte CHC, de Coimbra CAM, Machado JM, Guimarães ACD, Mendes KF, de Castro GC, Inoue MH (2023) Revisão Sobre Fatores Físicos e Químicos dos Herbicidas no Solo, Destacando o Conceito de Lixiviação e Seu Potencial. In: Atena (ed) Cultivando o futuro: tendências e desafios nas ciências agrárias 2, pp 68–84
94. Lipczynska-Kochany E (2018) Effect of climate change on humic substances and associated impacts on the quality of surface water and groundwater: a review. Sci Total Environ 640–641:1548–1565. https://doi.org/10.1016/J.SCITOTENV.2018.05.376
95. Liu J, Jiang J, Meng Y, Aihemaiti A, Xu Y, Xiang H, Gao Y, Chen X (2020) Preparation, environmental application and prospect of biochar-supported metal nanoparticles: a review. J Hazard Mater 388. https://doi.org/10.1016/j.jhazmat.2020.122026

96. Lokapur S, Gurikar R, Kulkarni GN (2014) Production and export of groundnut from India—an overview. Int Res J Agric Econ Stat 5(2):293–298. https://doi.org/10.15740/has/irjaes/5.2/293-298
97. Makepa DC, Chihobo CH, Musademba D (2023) Advances in sustainable biofuel production from fast pyrolysis of lignocellulosic biomass. Biofuels 14(5):529–550. https://doi.org/10.1080/17597269.2022.2151459
98. Mancuso MAC, Negrisoli E, Perim L (2011) Efeito residual de herbicidas no solo ("Carryover"). Revista Brasileira de Herbicidas 10(2):151–164. https://doi.org/10.7824/RBH.V10I2.106
99. Mašek O, Brownsort P, Cross A, Sohi S (2013) Influence of production conditions on the yield and environmental stability of biochar. Fuel 103:151–155. https://doi.org/10.1016/J.FUEL.2011.08.044
100. Medellín Castillo NA, González Fernández LA, Thiodjio-Sendja B, Aguilera-Flores MM, Leyva-Ramos R, Reyes-López SY, de León-Martínez LD, Dias JM (2023) Bone char for water treatment and environmental applications: a review. J Anal Appl Pyrol 175. https://doi.org/10.1016/j.jaap.2023.106161
101. Meftaul IM, Venkateswarlu K, Dharmarajan R, Annamalai P, Asaduzzaman M, Parven A, Megharaj M (2020) Controversies over human health and ecological impacts of glyphosate: is it to be banned in modern agriculture? Environ Pollut 263:114372. https://doi.org/10.1016/j.envpol.2020.114372
102. Mello FA, de Fagiani MAB, Silva RCR, Nai GA (2019) Agrotóxicos: impactos ao meio ambiente e à saúde humana. Colloquium Vitae 11(2):37–46. https://doi.org/10.5747/cv.2019.v11.n2.v262
103. Mendes KF, Hall KE, Takeshita V, Rossi ML, Tornisielo VL (2018) Animal bonechar increases sorption and decreases leaching potential of aminocyclopyrachlor and mesotrione in a tropical soil. Geoderma 316:11–18. https://doi.org/10.1016/j.geoderma.2017.12.017
104. Mendes KF, De Sousa RN, Takeshita V, Alonso FG, Régo APJ, Tornisielo VL (2019) Cow bone char as a sorbent to increase sorption and decrease mobility of hexazinone, metribuzin, and quinclorac in soil. Geoderma 343:40–49. https://doi.org/10.1016/j.geoderma.2019.02.009
105. Mendes KF, Furtado IF, Sousa RND, Lima ADC, Mielke KC, Brochado MGDS (2021) Cow bonechar decreases indaziflam pre-emergence herbicidal activity in tropical soil. J Environ Sci Health Part B 56:532–539. https://doi.org/10.1080/03601234.2021.1916302
106. Mendes KF, Freguglia RMO, Martins, BAB, Dias RC, Pimpinato RF, Tornisielo VL (2017) Cow bonechar for pesticide removal from drinking water. Scholars J Agric Vet Sci 4(11):504–512. https://doi.org/10.36347/sjavs.2017.v04i11.011
107. Meng L, Sun T, Li M, Saleem M, Zhang Q, Wang C (2019) Soil-applied biochar increases microbial diversity and wheat plant performance under herbicide fomesafen stress. Ecotoxicol Environ Saf 171:75–83. https://doi.org/10.1016/j.ecoenv.2018.12.065
108. Mielke KC, Castro GF, Mendes KF (2023) Cellulose acetate film containing bonechar for removal of metribuzin from contaminated drinking water. Processes 11(1). https://doi.org/10.3390/pr11010053
109. Mielke KC, Laube AFS, Guimarães T, da Brochado MGS, de Medeiros BAP, Mendes KF (2022) Pyrolysis temperature and application rate of sugarcane straw biochar influence sorption and desorption of metribuzin and soil chemical properties. Processes 10(10). https://doi.org/10.3390/pr10101924
110. Mielke K (2023) Sugarcane straw biochar and cow bonechar: physical-chemical characterization and interference on the behavior of metribuzin in soil and water. Universidade Federal de Viçosa, Tese
111. Mohan D, Sarswat A, Ok YS, Pittman CU (2014) Organic and inorganic contaminants removal from water with biochar, a renewable, low cost and sustainable adsorbent—a critical review. Bioresour Technol 160:191–202. https://doi.org/10.1016/J.BIORTECH.2014.01.120
112. Moussavi SP, Kadier A, Singh R, Ashoori R, Shirinkar M, Lu J, Zaidi NS, Sher F (2022) Superior removal of humic acid from aqueous stream using novel calf bones charcoal nanoadsorbent in a reversible process. Chemosphere 301. https://doi.org/10.1016/j.chemosphere.2022.134673

113. Mutchimadilok Y, Smittakorn S, Mongkolnchai-arunya S, Durnford D (2014) Defluoridation with locally produced Thai bone char. Adv Environ Chem 2014:1–9. https://doi.org/10.1155/2014/483609
114. Nair RR, Mondal MM, Weichgrebe D (2020) Biochar from co-pyrolysis of urban organic wastes—investigation of carbon sink potential using ATR-FTIR and TGA. Biomass Convers Biorefinery 12(10):4729–4743. https://doi.org/10.1007/S13399-020-01000-9
115. Nematian M, Ng'ombe JN, Keske C (2023) Sustaining agricultural economies: regional economic impacts of biochar production from waste orchard biomass in California's Central Valley. Environ Dev Sustain 1–21. https://doi.org/10.1007/S10668-023-03984-6/TABLES/4
116. Novak JM, Lima I, Xing B, Gaskin JW, Steiner C, Das KC, Ahmedna M, Rehrah D, Watts DW, Busscher WJ, Schomberg H (2009) Characterization of designer biochar produced at different temperatures and their effects on a loamy sand. Ann Environ Sci 3:195–206
117. Novotny EH, Auccaise R, Velloso MHR, Corrêa , Juliano Corulli, Higarashi MM, Abreu VMN, Rocha JD, kwapinski W (2012) Characterization of phosphate structures in biochar from swine bones. Pesqui Agropecu Bras 47(5):672–676
118. Oliveira DM (2021) Levantamento de dados de efeitos genotóxicos do herbicida glifosato e suas implicações cancerígenas em modelos animais. Universidade Federal de Uberlândia, Monografia
119. Oni BA, Oziegbe O, Olawole OO (2019) Significance of biochar application to the environment and economy. Ann Agricult Sci 64(2):222–236
120. Papageorgiou A, Sinha R, Azzi ES, Sundberg C, Enell A (2022) The role of biochar systems in the circular economy: biomass waste valorization and soil remediation. In: Zhang T (ed) The circular economy. IntechOpen, Rijeka
121. Patel S, Han J, Qiu W, Gao W (2015) Synthesis and characterisation of mesoporous bone char obtained by pyrolysis of animal bones, for environmental application. J Environ Chem Eng 3(4):2368–2377. https://doi.org/10.1016/j.jece.2015.07.031
122. Pauline AL, Joseph K (2020) Hydrothermal carbonization of organic wastes to carbonaceous solid fuel—a review of mechanisms and process parameters. Fuel 279:118472. https://doi.org/10.1016/J.FUEL.2020.118472
123. Paumgartten FJR (2020) Pesticides and public health in Brazil. Curr Opin Toxicol 22:7–11. https://doi.org/10.1016/J.COTOX.2020.01.003
124. Peter C, Helming K, Nendel C (2017) Do greenhouse gas emission calculations from energy crop cultivation reflect actual agricultural management practices?—A review of carbon footprint calculators. Renew Sustain Energy Rev 67:461–476
125. Piccirillo C (2023) Preparation, characterisation and applications of bone char, a food waste-derived sustainable material: a review. J Environ Manage 339:2023. https://doi.org/10.1016/j.jenvman.2023.117896
126. Piccirillo C (2023) Preparation, characterisation and applications of bone char, a food waste-derived sustainable material: a review. J Environ Manage 339:117896. https://doi.org/10.1016/J.JENVMAN.2023.117896
127. Prado-Acebo I, Cubero-Cardoso J, Lu-Chau TA, Eibes G (2024) Integral multi-valorization of agro-industrial wastes: a review. Waste Manage 183:42–52. https://doi.org/10.1016/j.wasman.2024.05.001
128. Qin L, He L, Yang W, Lin A (2020) Preparation of a novel iron-based biochar composite for removal of hexavalent chromium in water. Environ Sci Pollut Res 27(9):9214–9226. https://doi.org/10.1007/s11356-019-06954-6
129. Quintana-Najera J, Blacker AJ, Fletcher LA, Ross AB (2021) The effect of augmentation of biochar and hydrochar in anaerobic digestion of a model substrate. Bioresour Technol 321. https://doi.org/10.1016/j.biortech.2020.124494
130. Quispe JIB, Campos LC, Mašek O, Bogush A (2022) Use of biochar-based column filtration systems for greywater treatment: a systematic literature review. J Water Process Eng 48
131. Rani L, Thapa K, Kanojia N, Sharma N, Singh S, Grewal AS, Srivastav AL, Kaushal J (2021) An extensive review on the consequences of chemical pesticides on human health and environment. J Clean Prod 283:124657. https://doi.org/10.1016/J.JCLEPRO.2020.124657

132. Rodriguez JA, Lustosa Filho JF, Melo LCA, de Assis IR, de Oliveira TS (2020) Influence of pyrolysis temperature and feedstock on the properties of biochars produced from agricultural and industrial wastes. J Anal Appl Pyrol 149. https://doi.org/10.1016/j.jaap.2020.104839
133. Sadh PK, Duhan S, Duhan JS (2018) Agro-industrial wastes and their utilization using solid state fermentation: a review. Bioresour Bioprocess 5(1):1–15. https://doi.org/10.1186/s40643-017-0187-z
134. Sagar PR, Raol GG, Prajapati D, Kapdi D, Kiri B (2024) Enhanced productivity of nutrient-rich single cell protein from Paradendryphiella arenariae PG1 through valorization of agro-industrial waste: a Green symphony from waste-to-protein approach. Bioresour Technol Rep 26. https://doi.org/10.1016/j.biteb.2024.101884
135. Sakhiya AK, Baghel P, Pathak S, Vijay VK, Kaushal P (2020) Effect of process parameters on slow pyrolysis of rice straw: product yield and energy analysis. In: Proceedings of the 2020 international conference and utility exhibition on energy, environment and climate change, ICUE 2020. Institute of Electrical and Electronics Engineers Inc.
136. Sakhiya AK, Vijay VK, Kaushal P (2023) Development of rice straw biochar through pyrolysis to improve drinking water quality in arsenic and manganese contaminated areas. Surf Interfaces 36. https://doi.org/10.1016/j.surfin.2022.102582
137. Santos C, Mancuso M, Toebe M, Schuller T (2021) Mapeamento do Potencial de Contaminação das Águas Subterrâneas e Superficiais por Agrotóxicos no Noroeste Do Rio Grande Do Sul. Revista Geociências 40(4):951–966
138. Santos JA, Gonzaga MIS, de Almeida AQ, da Silva AJ, de Santos JCJ, da Lima IS (2022) Caracterização, retenção e disponibilidade de água de diferentes tipos de biocarvão de origem animal e vegetal. Res Soc Dev 11(5):e48411528360. https://doi.org/10.33448/rsd-v11i5.28360
139. Schmidt HP, Bucheli T, Kammann C, Abiven S, Leifeld J, Soja G, Hagemann N (2022) European biochar certificate—guidelines for a sustainable production of biochar. European Biochar Foundation 10.0
140. Setter C, Silva FTM, Assis MR, Ataíde CH, Trugilho PF, Oliveira TJP (2020) Slow pyrolysis of coffee husk briquettes: characterization of the solid and liquid fractions. Fuel 261. https://doi.org/10.1016/j.fuel.2019.116420
141. Sharma R (2024) Agro-industrial waste to energy—sustainable management. Sustain Mater Technol 41
142. Shepherd JG, Sohi SP, Heal KV (2016) Optimising the recovery and re-use of phosphorus from wastewater effluent for sustainable fertiliser development. Water Res 94:155–165. https://doi.org/10.1016/J.WATRES.2016.02.038
143. Silva MCS (2020) Investigação da ocorrência de resíduos de herbicidas em rios da reserva extrativista Acaú-Goiana. Universidade Federal de Pernambuco, Tesa
144. Singh E, Mishra R, Kumar A, Shukla SK, Lo S-L, Kumar S (2022) Circular economy-based environmental management using biochar: driving towards sustainability. Process Saf Environ Prot 163:585–600. https://doi.org/10.1016/j.psep.2022.05.056
145. Singh Y, Sharma S, Kumar U, Sihag P, Balyan P, Singh KP, Dhankher OP (2024) Strategies for economic utilization of rice straw residues into value-added by-products and prevention of environmental pollution. Sci Total Environ 906
146. Singha A, Rahaman MdA, Jiku MdAS, Faruquee M, Alam MdA, Sinha SR (2019) Biochar application and no-tillage practices to minimize the residues of herbicides in the seeding hole. Bull Natl Res Cent 43:165. https://doi.org/10.1186/s42269-019-0222-7
147. Smernik RJ (2009) Biochar and sorption of organic compounds. In: Earthscan (ed) Biochar for environmental management. Reino Unido, pp 289–300
148. Soares MB, dos Santos FH, Alleoni LRF (2021) Iron-modified biochar from sugarcane straw to remove arsenic and lead from contaminated water. Water Air Soil Pollut 232(9). https://doi.org/10.1007/s11270-021-05343-5
149. Song W, Guo M (2012) Quality variations of poultry litter biochar generated at different pyrolysis temperatures. J Anal Appl Pyrol 94:138–145. https://doi.org/10.1016/j.jaap.2011.11.018

150. Song J, Wang Y, Zhang S, Song Y, Xue S, Liu L, Lvy X, Wang X, Yang G (2021) Coupling biochar with anaerobic digestion in a circular economy perspective: a promising way to promote sustainable energy, environment and agriculture development in China. Renew Sustain Energy Rev 144:110973. https://doi.org/10.1016/j.rser.2021.110973
151. Sun J, He F, Pan Y, Zhang Z (2017) Effects of pyrolysis temperature and residence time on physicochemical properties of different biochar types. Acta Agric Scand B Soil Plant Sci 67(1):12–22. https://doi.org/10.1080/09064710.2016.1214745/ASSET/789D8BBA-542C-47B0-9C26-B6E4AC6CE0B1/ASSETS/IMAGES/SAGB_A_1214745_F0008_B.GIF
152. Suresh Babu KKB, Nataraj M, Tayappa M, Vyas Y, Mishra RK, Acharya B (2024) Production of biochar from waste biomass using slow pyrolysis: studies of the effect of pyrolysis temperature and holding time on biochar yield and properties. Mater Sci Energy Technol 7:318–334. https://doi.org/10.1016/j.mset.2024.05.002
153. Száková J, Stiborová H, Mercl F, Hailegnaw NS, Lhotka M, Derevyankina T, Paul CS, Taisheva A, Brabec M, Tlustoš P (2023) Woodchips biochar versus bone char in a one-year model soil incubation experiment: the importance of soil/char pH alteration on nutrient availability in soil. J Chem Technol Biotechnol. https://doi.org/10.1002/JCTB.7421
154. Taskin E, Branà MT, Altomare C, Loffredo E (2019) Biochar and hydrochar from waste biomass promote the growth and enzyme activity of soil-resident ligninolytic fungi. Heliyon 5:e02051. https://doi.org/10.1016/J.HELIYON.2019.E02051
155. Thakkar J, Kumar A, Ghatora S, Canter C (2016) Energy balance and greenhouse gas emissions from the production and sequestration of charcoal from agricultural residues. Renew Energy 94:558–567. https://doi.org/10.1016/j.renene.2016.03.087
156. Thangarajan R, Bolan NS, Tian G, Naidu R, Kunhikrishnan A (2013) Role of organic amendment application on greenhouse gas emission from soil. Sci Total Environ 465:72–96
157. Thilagavathi T, Janaki P, Kumar RM (2023) Enhancing herbicides sorption in sodic soils through biochar amendment : a promising approach for sustainable agriculture. JANS 15:1445–1452. https://doi.org/10.31018/jans.v15i4.5008
158. Tolera ST, Alemu FK (2020) Potential of abattoir waste for bioenergy as sustainable management, eastern Ethiopia, 2019. J Energy 2020:1–9. https://doi.org/10.1155/2020/6761328
159. Tomczyk A, Sokołowska Z, Boguta P (2020) Biochar physicochemical properties: pyrolysis temperature and feedstock kind effects. Rev Environ Sci Biotechnol 19(1):191–215
160. Tripathi M, Sahu JN, Ganesan P (2016) Effect of process parameters on production of biochar from biomass waste through pyrolysis: a review. Renew Sustain Energy Rev 55:467–481
161. U.S. Department of Agriculture (USDA). Production commodity data. https://fas.usda.gov/data/production/commodity/0711100. Acesso em: 07 out. 2024
162. Ucella-Filho JGM, Ferreira NS, Lacerda EMF, Soriano Neto C, de Duarte CES, Ferreira IM, Balu AM, Ignacchiti MDC, Dias Junior AF, Resende JA (2024) Citrus sinensis waste wood pyroligneous extract as an effective agent against biofilms formed by mastitis-causing Staphylococcus aureus. Ind Crops Prod 218. https://doi.org/10.1016/j.indcrop.2024.118854
163. Uchimiya M, Wartelle LH, Lima IM, Klasson KT (2010) Sorption of deisopropylatrazine on broiler litter biochars. J Agric Food Chem 58:12350–12356. https://doi.org/10.1021/jf102152q
164. United Nations Department of Economic and Social (2022) World population prospects 2022: summary of results
165. United Nations (2024) Objetivos de desenvolvimento sustentável. https://sdgs.un.org/es/goals
166. United Nations (2015) Transforming our world: the 2030 agenda for sustainable development. United Nations, New York. Available at: https://sdgs.un.org/2030agenda. Accessed 16 Apr 2025
167. Verduzo Garibay M, Fernández del Castillo A, de Anda J, Senés-Guerrero C, Gradilla-Hernández MS (2021) Structure and activity of microbial communities in response to environmental, operational, and design factors in constructed wetlands. Int J Environ Sci Technol 19(11):11587–11612. https://doi.org/10.1007/S13762-021-03719-Y

168. Wakweya T, Nigussie A, Worku G, Biresaw A, Aticho A, Hirko O, Ambaw G, Mamuye M, Dume B, Ahmed M (2022) Long-term effects of bone char and lignocellulosic biochar-based soil amendments on phosphorus adsorption–desorption and crop yield in low-input acidic soils. Soil Use Manage 38(1):703–713. https://doi.org/10.1111/sum.12757
169. Wang Z, Liu K, Xie L, Zhu H, Ji S, Shu X (2019) Effects of residence time on characteristics of biochars prepared via co-pyrolysis of sewage sludge and cotton stalks. J Anal Appl Pyrol 142:104659. https://doi.org/10.1016/J.JAAP.2019.104659
170. Wang J, Wang S (2019) Preparation, modification and environmental application of biochar: a review. J Clean Prod 227:1002–1022. https://doi.org/10.1016/j.jclepro.2019.04.282
171. Wang D, Jiang P, Zhang H, Yuan W (2020) Biochar production and applications in agro and forestry systems: a review. Sci Total Environ 723. https://doi.org/10.1016/j.scitotenv.2020.137775
172. Wang D, Mukome FND, Yan D, Wang H, Scow KM, Parikh SJ (2015) Phenylurea herbicide sorption to biochars and agricultural soil. J Environ Sci Health Part B 50:544–551. https://doi.org/10.1080/03601234.2015.1028830
173. Water UN et al (2018) Sustainable development goals a SRHR CSO guide for national implementation
174. Weber K, Quicker P (2018) Properties of biochar. Fuel 217:240–261. https://doi.org/10.1016/j.fuel.2017.12.054
175. Xiao J, Huang J, Huang M, Chen M, Wang M (2020) Application of basalt fiber in vertical flow constructed wetland for different pollution loads wastewater: performance, substrate enzyme activity and microorganism community. Bioresour Technol 318:124229. https://doi.org/10.1016/J.BIORTECH.2020.124229
176. Yaashikaa PR, Kumar PS, Varjani S, Saravanan A (2020) A critical review on the biochar production techniques, characterization, stability and applications for circular bioeconomy. Biotechnology Reports 28:e00570. https://doi.org/10.1016/j.btre.2020.e00570
177. Yaashikaa PR, Kumar PS, Varjani S, Saravanan A (2020) A critical review on the biochar production techniques, characterization, stability and applications for circular bioeconomy. Biotechnology Reports 28. https://doi.org/10.1016/j.btre.2020.e00570
178. Zango ZU, Garba A, Haruna A, Imam SS, Katsina AU, Ali AF, Abidin AZ, Zango MU, Garba ZN, Hosseini-Bandegharaei A, Yuguda AU, Adamu H (2024) A systematic review on applications of biochar and activated carbon derived from biomass as adsorbents for sustainable remediation of antibiotics from pharmaceutical wastewater. J Water Process Eng 67
179. Zhang X, Zhao B, Liu H, Zhao Y, Li L (2022) Effects of pyrolysis temperature on biochar's characteristics and speciation and environmental risks of heavy metals in sewage sludge biochars. Environ Technol Innov 26:102288. https://doi.org/10.1016/J.ETI.2022.102288
180. Zhang Y, Wang J, Feng Y (2021) The effects of biochar addition on soil physicochemical properties: a review. Catena (Amst) 202:105284. https://doi.org/10.1016/J.CATENA.2021.105284
181. Zhu Y, Dai W, Deng K, Pan T, Guan Z (2020) Efficient removal of Cr(VI) from aqueous solution by Fe-Mn oxide-modified biochar. Water Air Soil Pollut 231(2). https://doi.org/10.1007/s11270-020-4432-2
182. Zwetsloot MJ, Lehmann J, Solomon D (2015) Recycling slaughterhouse waste into fertilizer: how do pyrolysis temperature and biomass additions affect phosphorus availability and chemistry? J Sci Food Agric 95(2):281–288. https://doi.org/10.1002/jsfa.6716

Açaí Waste as Raw Material for Bioactive Compounds: A Social Context of the Production Chain and Pathways for Sustainable Production in the Amazon Rainforest

Allana Katiussya Silva Pereira, João Paulo Bispo Santos, Paulo Renato Souza de Oliveira, Irlon Maciel Ferreira, Alex Bruno Lobato Rodrigues, and Ananias Francisco Dias Júnior

Abstract The açaí fruit, harvested from the *Euterpe oleracea* Mart. and *Euterpe precatoria* Mart. palm trees, is native to the Amazon region of Brazil and plays a crucial role in the socioeconomic and cultural lives of indigenous and riverine communities. Deeply embedded in the traditional practices of these communities, açaí serves as both a staple food source and a primary means of livelihood. Over the past decade, the growing global demand for nutritious and health-promoting products

A. K. S. Pereira (✉) · J. P. B. Santos · P. R. S. de Oliveira
Department of Forest Sciences, "Luiz de Queiroz" College of Agriculture (Esalq), Universidade de São Paulo (USP), Av. Pádua Dias, 11, Piracicaba, SP 13418-900, Brazil
e-mail: allana.florestal@gmail.com

J. P. B. Santos
e-mail: bisposjp@gmail.com

P. R. S. de Oliveira
e-mail: pr.oliveira0@gmail.com

J. P. B. Santos
Department of Biology, Universidade Estadual de Campinas (Unicamp), University City Zeferino Vaz, Campinas, SP 13083-970, Brazil

P. R. S. de Oliveira · A. F. Dias Júnior
Agricultural Sciences and Engineering Center, Universidade Federal do Espírito Santo (Ufes), Av. Governador Lindemberg, 316, Jerônimo Monteiro, ES 29550-000, Brazil
e-mail: ananias.dias@ufes.br

I. M. Ferreira · A. B. L. Rodrigues
Biocatalysis and Applied Organic Synthesis Laboratory, Universidade Federal do Amapá (Unifap), Campus Marco Zero do Equador, Rodovia Josmar Chaves Pinto, Km 02, Bairro Zerão, Macapá, AP 68903-419, Brazil
e-mail: irlon.ferreira@unifap.br

A. B. L. Rodrigues
e-mail: alex.rodrigues@unifap.br

A. K. Silva Pereira and A. F. Dias Júnior (eds.), *Agro-industrial Waste for New Pharmaceuticals*, Green Energy and Technology,
https://doi.org/10.1007/978-3-031-88534-1_4

has caused açaí consumption to more than double, establishing it as the most valuable crop in Brazil's plant-based food sector. This trend continues as global açaí consumption expands, with market analysts estimating the global açaí market at $1.23 billion in 2024, rising to $3.09 billion by 2032. However, the rapid expansion of the açaí market has introduced considerable challenges in waste management, specifically regarding the disposal of seeds and fibers generated during fruit pulping. Ensuring the sustainability of açaí production requires viewing this waste as a valuable resource for high-value product manufacturing. Research has indicated that residues from the açaí production chain possess significant bioactive properties, including antioxidant, anti-inflammatory, and biotechnological applications, making them valuable resources for a variety of industries. These by-products also represent a promising foundation for developing therapeutic applications capable of fostering technological innovation and job creation. To advance sustainable development discussions within the açaí production chain, this chapter explores strategies for valorizing açaí waste, with a focus on its bioactive properties and potential applications. The insights provided here aim to contribute to the development of novel therapeutic products with potential benefits for public health, ultimately supporting the sustainable expansion of the açaí industry.

Keywords Açaí palm · Amazonian fruit · Reuse waste · Bioactive compounds · Sustainable bioeconomy

1 Introduction

The açaí fruit, derived from the açaí palm (*Euterpe oleracea* Mart. and *Euterpe precatoria* Mart.), native to several countries in the Amazon region, is an endemic fruit that plays a pivotal role for indigenous and riverine communities, where its harvesting and consumption is an important source of income [24]. Beyond its economic importance, the açaí fruit is also deeply embedded in the complex socio-biocultural context of Amazonian communities, comprising an essential part of their cultural and traditional practices [24, 42]. Due to its wide range of applications, from food products (e.g., açaí wine and pulp) to medicinal uses by traditional people (e.g., bioactive compounds with antioxidant properties), the fruit can be consumed in numerous forms [42]. The Brazilian Institute of Geography and Statistics [36] estimated that approximately 1.7 million tons of açaí were produced in Brazil in 2023, this is an abundant generation of waste that needs to be managed and recovered. For every kilogram of açaí produced, 2–4 kg of waste are generated [73]. With an annual growth rate projected at 12.2% for the açaí market between 2024 and 2032 [49], the waste generated is also expected to increase proportionally.

The global demand for healthy products has increased the demand for açaí, because in addition to its properties as a "superfood"—one that not only nourishes but also provides additional health benefits—the value of açaí goes beyond its nutritional and cultural significance [63]. As a result, demand for açaí fruit has doubled in the last decade, from 116,000 tons in 2009 to more than 227,00 tons in 2021 [75]. In recent decades, this Amazonian fruit has gained significant attention in the international market and is rapidly expanding as a functional food valued for health benefits [5, 75]. The açaí fruit has been shown to have important biological activities that are beneficial to human health due to its rich fruit composition that includes antioxidant elements [9] anti-inflammatory properties [47] and cancer prevention [35].

The rich chemical composition of açaí allows the development of several new products with different formulations, which has a strong impact on the generation of income, jobs, and new businesses, and stimulates the development of research and innovation [47]. The high concentration of bioactive compounds in açaí pulp has led industries and research centers to investigate and invest in the development of new products in various industries, such as cosmetics, food, and medicine [45]. However, these bioactive properties are not limited to açaí pulp alone: the production chain residues, particularly seeds and fibers, are potential raw materials for various products and exhibit bioactive properties similar to those of the pulp. These materials have been studied as sources of lipids, fibers, fatty acids, sterols, and phenolic compounds with antioxidant activity [11, 17, 54, 86].

Açaí seeds represent about 85% of the total fruit, which is often discarded in waste dumps without specific management [6, 25]. Recent studies have shown that approximately 53% of açaí residues are inadequately disposed of, mainly due to a lack of environmental legislation and specific management practices [25]. These raw materials could serve as valuable sources of bioactive compounds for use in the medical or cosmetic industry (e.g., antioxidant cosmetics and medicinal oils) or be reused in manufactured products (e.g., biodegradable packaging, building materials, and textiles), thus reducing their disposal in landfills [6].

Açaí production has significant negative social and environmental impacts due to the large quantities of residues that are inadequately disposed of [6]. The wastes generated after açaí pulping, often accumulate in landfills due to a lack of alternative uses [62]. This situation raises concerns about the efficiency and environmental impact of the açaí supply chain. Consequently, the problem of waste has a negative impact on the production chain, which hinders industrial development and the practice of a more sustainable bioeconomy, leading to market instability, economic sanctions and environmental fines [6]. Discussions that support the management and valorization of waste from agricultural and forestry activities are essential for reducing environmental impacts and promoting the adoption of alternative technologies. When referring to açaí, the emblematic fruit of the Western Amazon, we are also addressing a production chain characterized by low yields in products and high yields in waste.

Açaí waste offer a wide range of application possibilities and have paved the way for innovative scientific studies and research and development programs focused on potential uses, while simultaneously creating new economic opportunities and addressing sustainability issues [6]. Recently, several studies have investigated the use and applications of açaí waste [38, 66, 68]. Igreja et al. [38] highlighted that açaí seeds are a valuable agroindustrial by-product with great potential as a low-cost substrate for biotechnological applications, offering environmental benefits but also supporting bioeconomic development. A recent study by Sato et al. [68] demonstrated that açaí seeds can be converted into biochar, which has significant potential for soil carbon sequestration [40, 68]. According to Sato et al. [68], this innovative application can not only contribute to long-term soil carbon storage, but also improve soil health by regulating acidity and enhancing fertility, promoting sustainable agricultural practices. Other studies have demonstrated the management of açaí waste and its potential for bioenergy production [43, 44, 33, 71].

In terms of medicinal applications, the residual biomass of açaí retains a considerable amount of bioactive compounds, such as phenolics (i.e. antioxidants), which have a high potential for applications in various industrial sectors, including medicine [13]. Several studies have shown that these compounds, especially flavonoids and anthocyanins, are present in high concentrations in the waste from açaí pulp production and that it is possible to extract them for alternative uses and applications [10, 13, 14]. In addition, a recent literature review suggests that açaí is safe and shows promise as a chemoprotective agent against cancer [2]. Its rich composition of bioactive compounds, such as anthocyanins and flavonoids, has been linked to the inhibition of cancer cell growth and the reduction of oxidative stress, which plays a key role in cancer development [2, 35], with anticarcinogenic effects observed in seed extracts from the reduction in tumor diameter and weight analyzed [31]. These findings suggest that waste açaí-based therapies may represent a novel and complementary strategy for cancer treatment, offering both preventive and therapeutic benefits in oncology.

These applications not only add value to açaí waste, but also support a circular economy by reducing environmental impact and creating high-value products. Once a vital part of Amazonian communities, the açaí fruit has gained global popularity for its health benefits and bioactive compounds [85]. By converting açaí residues into valuable products such as biochar, bioenergy and bioactive ingredients for pharmaceuticals and cosmetics, the industry can reduce its environmental footprint while fostering innovation. Thus, in this chapter we address everything from the historical and sociocultural aspects involved in the açaí production chain to the valorization of its residues through the use of their bioactive potential. This approach opens up new economic opportunities and promotes a sustainable bioeconomy, ensuring responsible growth of the açaí market in harmony with the environment.

2 History and Sociocultural Context of Açaí Production in South America

The Amazon region is vast, covering an area of approximately 6.74 million km^2, more than 60% of which is in Brazil. This region also extends through Bolivia, Colombia, Ecuador, Guyana, Peru, Venezuela, Suriname, and French Guiana. If it were considered a country, the Amazon would be the seventh largest nation in the world in terms of area. Given this territorial extent, we find a diversity that transcends geographical boundaries, encompassing a multitude of peoples, cultures, accents, and of course a hiper diversity of fauna and flora. Among its native ecosystems, the most well-known is the tropical rainforest, which plays a central role in global climate regulation due to its biodiversity and carbon storage capacity.

In this territorial context, two frameworks are crucial for the sustainable maintenance and development of the region, interconnected and interdependent. The first is sociobiodiversity, a concept that encompasses the relationship between biological diversity, traditional agricultural systems, and natural resource management, integrating the knowledge and culture of traditional communities and small farmers. The second is the bioeconomy, a concept originally proposed by Georgescu-Roeggen, that challenges the logic of perpetual economic growth and focuses on the responsible use of resources to meet the needs of the most vulnerable populations [34]. Together, sociobiodiversity and the bioeconomy offer a promising way to reconcile environmental conservation with economic and social goals. However, for this reconciliation to be effective, however, it is essential to integrate biodiversity more robustly into bioeconomic approaches. In traditional practices, this integration is still limited, and there is an urgent need to place it at the center of sustainability strategies. Furthermore, issues of resource distribution, equity, and social responsibility must be addressed for the bioeconomy to become an effective and fair tool. The açaí palm represents an important connection between these two concepts. Historically, it has been part of the life of Amazonian population, while at the same time being a high-value economic product. Thus, the açaí palm not only strengthens the region's cultural identity but is also an essential resource for the development of a sustainable bio-economy.

The açaí palm has a slender and elegant shape and developsin floodplains and along the banks of upland rivers (Fig. 1). As a characteristic forest species of the region, its perennial nature makes it suitable for tropical conditions, characterized by high rainfall and elevated temperatures, providing continuous coverage [15].

There are two species of açaí palm. The first is *Euterpe oleracea* (Mart), commonly known as cluster açaí,[1] whose natural occurrence is concentrated in the eastern Amazon, considered its center of origin and genetic diversity. It grows spontaneously in the states of Amapá, Maranhão, Pará, and Tocantins, and can also be found in Mato Grosso, forming extensive colonies near streams and rivers [83] (Fig. 2). In this region, dense and diverse populations can be found, especially in areas subject to periodic flooding due to the ebb and flow of the tides.

The second species is *Euterpe precatoria* (Mart), known as solitary açaí, which predominantly occurs primarily in the western Amazon [64], in low-altitude tropical forests, usually near water bodies in periodically flooded areas in the states of Acre, Amazonas, Pará, and Rondônia [84] (Fig. 3).

In the Brazilian territory, three cultivars of *E. oleracea* have a place in agriculture: Estuário, BRS Pará, and BRS Pai d'Égua. All of them were developed through the efforts and research of the Empresa Brasileira de Pesquisa Agropecuária [Brazilian

[1] Clusters are sprouts that naturally occur in cespitose habit palms. For *Euterpe oleracea*, the sprouts are at different stages of maturation and can be numerous, reaching up to 25 individuals per cluster.

Fig. 1 Overview of Açaí palm. *Source* Authors (2024)

Agricultural Research Corporation] (Embrapa) [46] and adapt to different cultivation strategies or objectives. Among the notable physical characteristics common to all varieties of *E. oleracea* are the fruits, which are globose drupes with a diameter of 1–2 cm and an average weight of 1.5 g, ranging from 0.5 to 2.8 g. The exocarp, a thin outer layer and the edible part of the fruit, varies from purple to grayish-green when ripe, depending on the type of açaí. The mesocarp, which is almost imperceptible, consists of a thin fleshy layer about 1 mm thick, which is also edible and surrounds the voluminous, rigid, and fibrous endocarp on the outside. This endocarp follows the shape of the fruit and houses the seed. Popularly, what is referred to as the seed is actually the pyrene, since the seed is protected by the endocarp [16, 29].

The most traditional uses of the açaí palm include the consumption of its fruits and the heart of the palm, extending to the use of its trunks as construction material by indigenous populations [3]. The leaves can also be used as roofing for buildings, and the seeds serve as natural fertilizers [15]. The natural order of the palm is to grow until it reaches maturity, at which point it produces fruit. When mature, individuals can be cut down for the heart of the palm extraction, with the trunk and leaves then used as building materials. It is not uncommon for areas near traditional communities to be filled with açaí palms. The harvest of its products is done through main trails that branch into smaller trails that connect different gathering areas [1]. Studies support the hypothesis that the species associated with what we call the açaí palm are hyper

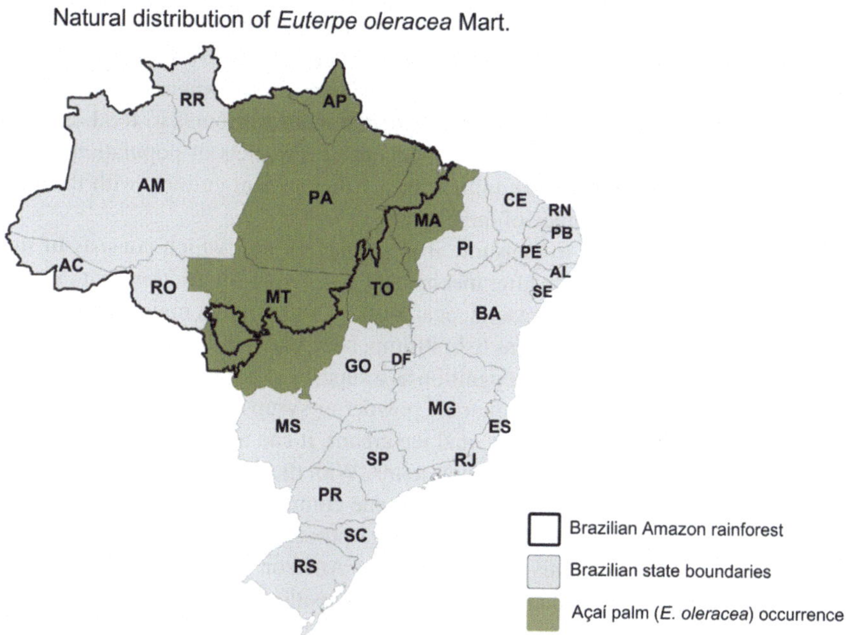

Fig. 2 Natural occurrence of *Euterpe oleracea* Mart. *Source* Authors (2024)

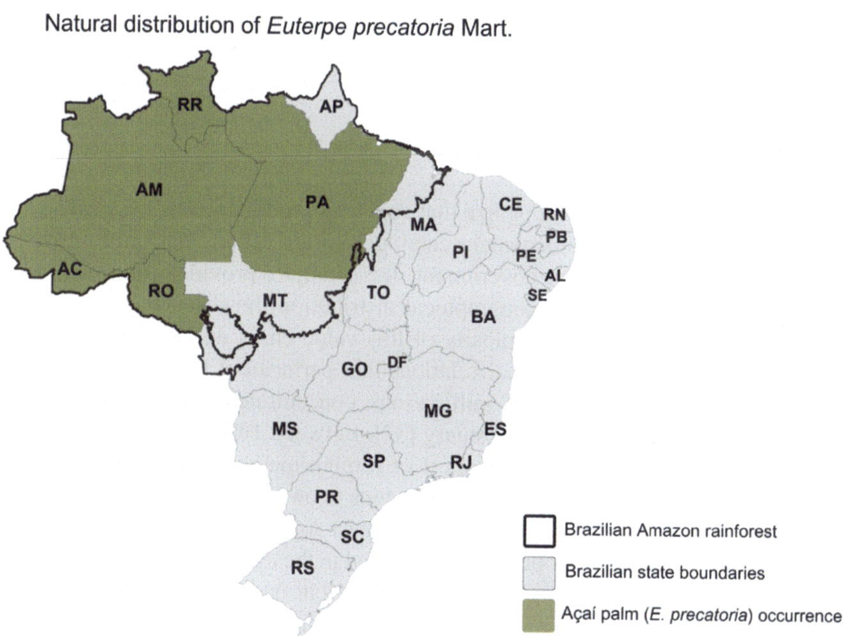

Fig. 3 Natural occurrence of *Euterpe precatoria* Mart. *Source* Authors (2024)

dominant throughout the Amazon basin, due to anthropogenic factors provided by indigenous populations over the last five to ten millennia [57, 79]. Furthermore, the presence of this palm extends to home gardens and agroforests, where it is an integral part of a complex multifaceted agricultural system that contributes to food security and income for rural producers [70]. In this context, the diets of populations from different parts of the Amazon are influenced by this ancient culture, with the açaí fruit often being the centerpiece of their meals.

The most common way to consume it is as "açaí wine," which consists of the product extracted from the fruit after mechanical processing (by friction) with water. It is worth noting that, in the Amazon, açaí wine is considered a main dish, usually accompanied by other foods such as fish, shrimp, beef, cassava flour, or tapioca flour. In other regions, açaí is gaining recognition as a superfood with energetic properties, consumed after being mixed with guaraná powder or syrup and other additives, and used as a base for juices, smoothies, and ice cream. It can also be accompanied by fruits, seeds, condensed milk, and similar items. With the growth of the fruit market and advances in the food industry, it has become common to find various products based on açaí pulp, including processed and ultra-processed items [75]. Unlike a few years ago, when açaí was mainly found outside the Amazon in the form of frozen pulp or powdered products [56], today we see a clear transition. Once part of Amazonian tradition and culture as food and a source of income for local populations, açaí is now also consolidating as a global commodity. In this market, it is primarily valued as a functional food, attracting consumers interested in its health benefits, especially for its antioxidant and energetic properties.

3 The Açaí Production Chain

In general, it is important to recognize that açaí fruit production has regional specificities that are not always adequately cataloged in the official data from government agencies. The Brazilian government, for example, provides information on açaí production volumes with geographical distribution; however, these databases are limited in reflecting the complexity of this value chain due to the high informality of the commercial and labor relations that permeate it [64]. Last year, açaí production in Brazil was about 1.7 million tons, concentrated mainly in the state of Pará, which is part of the Amazon estuary [37]. Between 1987 and 2022, there was significant growth driven by the conquest of foreign markets and other regions of Brazil, the production jumped from 145,800 tons to an impressive 1.9 million tons, an increase of over 13 times in 36 years. The latest Agricultural Census, conducted in 2017, shows that 81,000 properties are involved in the extraction of açaí fruit [39]. Most of these properties do not cultivate the palm but operate within an extraction or management system. A production chain based on native biomes predominates, managed by traditional populations, and generally focused on the local consumer market. We will explore this chain further below.

In natural environments, the açaí palm is abundant. The species E. oleracea also exhibits an oligarchic character and has a significant representation within a plant community, reaching up to 397 individuals per hectare [79]. As summarized by Quaresma and Euler [64], the açaí production chain is a link between (1) extractivist communities that collect the fruit either independently or organized in cooperatives, (2) intermediaries that handle purchasing and transportation, (3) markets in ports and commercial points, (4) agro-industries that process the fruit into açaí wine for local consumption or resell it for the production of other products,and (5) the final consumer. Groups of extractivists tend to increase their power of action and negotiation at different links of the production chain [64], which is further strengthened by the absence of intermediaries. The management of native açaí areas is a key point for the conservation of diversity and economic activity. Embrapa's studies provide basic guidelines that promote proper use and have been an effective tool in the fruit chain. Unmanaged areas can produce between 1 and 2 tons per hectare per year, while managed areas reach an average of 5–6 tons per hectare per year [65].

The traditional harvesting of açaí has been previously described by Oliveira et al. [28]. It is typically performed early in the morning, with men climbing the palm trunk using a special tool known as a "peconha", along with a knife to cut the bunches. Modern equipment, such as adjustable aluminum poles and pulley devices, is also being successfully tested. After harvesting, the fruit must be removed from the bunches, cleaned, and placed in ventilated baskets or plastic boxes. Due to its perishable nature, the fruits must be processed within 24 h, avoiding exposure to sunlight. For long distance transportation, the use of polypropylene bags with ice or cold chambers is recommended. Processing occurs traditionally, where the washed fruits are pulped in water and sifted using instruments made of plant fibers. Furthermore, it is most common to carry out the depulping process mechanically [58].

This equipment consists of a stainless-steel cylinder in which the fruit is pressed by blades. The cylinder, which is vertically positioned, has a mobile central shaft with three plano-convex blades arranged perpendicularly, forming a cross that covers the entire diameter of the cylinder. The first blade is at the base of the shaft, and the other two are spaced apart, with the intermediate blade positioned at a 90° angle to the others. The fruit is placed manually at the top of the cylinder, with a small amount of water added to facilitate the drainage of the pulp. After passing through a sieve, the pulp drains like açaí wine. The depulped seeds are removed by gravity through an opening at the bottom of the machine and are considered waste. It is important to note that in urban and rural areas of the Amazon, small and micro agro-industries use depulping machines to separate the wine from the waste and commercialize the açaí. The seeds are accumulated in storage facilities or in places near commercial establishments, becoming an environmental liability [12].

In the açaí production chain, the seed (Fig. 4) corresponds to the endocarp and kernel of the fruit [17]. It is a rich organic material that has attracted the interest of many researchers, particularly for its reuse in energy production [43, 33, 45, 72]. Due to its high lignin content, the açaí seed is considered to decompose slowly, unlike other plant materials, such as açaí leaves.

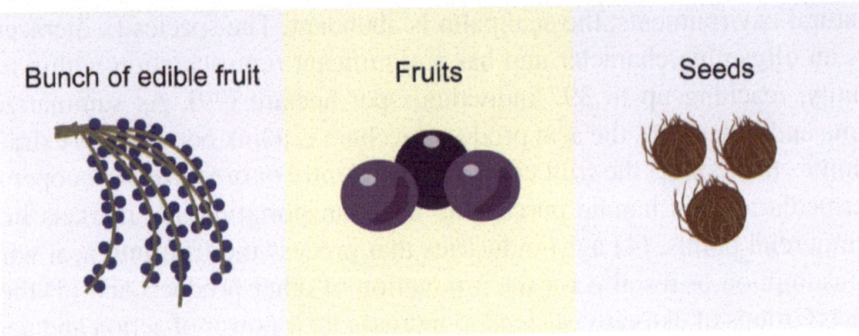

Fig. 4 Açaí fruit and seed. *Source* Authors (2024)

The açaí seed, which lacks the exocarp and part of the mesocarp, is composed of an embryonic axis and abundant endosperm tissue, representing 74.6–80.9% of the fruit's mass [61], with reports in the literature of up to 85% [6]. The seeds can be subjected to forced drying to segregate the fibrous mass from the defibrated seed [32]. Açaí fibers can be a potentially useful raw material for producing medium-density homogeneous boards for application in civil construction and the furniture industry [26]. During the pulping process, the pulp is separated from the seeds (which constitute the first fraction of the waste) (Fig. 5). In a second step, the pulp is clarified through a filter. At this stage, a paste made up of fibers and other solid waste produced during the separation of the pulp from the seeds is removed (thus forming a second fraction of the waste) [13]. The wine or edible pulp is the most significant coproduct of the açaí processing chain as it is widely used for consumption *in natura* in the food industry and as a raw material for natural dyes [8]. Another interesting product of açaí fruits is the oil. Through a mechanical extraction process performed by cold pressing, this product is obtained and is widely used in the cosmetics industry (accounting for 19% of the açaí fruit). The extraction typically occurs through cold pressing of freeze-dried pulp, preserving biological and physicochemical properties. This process is viable for industrial scale, although it yields low outputs [6]. Another residue from the açaí production chain mentioned in the literature is the residual cake resulting from this cold pressing, which accounts for 11% of the fruit [74].

4 Valorization of Açaí Waste: Bioactive Potential and Sustainable Applications

Plant bioactive compounds, also known as phytochemicals, are extra-nutritional substances found in small amounts in foods that offer health benefits. Although these compounds are not essential for nutrition, they play significant roles in the prevention and treatment of chronic diseases, such as cardiovascular diseases and cancer, due to their antioxidant, anti-inflammatory, and anticancer properties. The efficacy of these secondary metabolites, including polyphenols, flavonoids, and carotenoids,

Fig. 5 Approximate yields of açaí products and waste found in the literature. *Source* Authors (2024)

depends on factors like the ingested dose and their mechanisms of action in metabolic and cellular processes, underscoring their positive impact on human health [78]. A notable example is açaí which is rich in bioactive compounds of scientific interest.

Açaí has a rich and diverse chemical composition, making it a promising source of bioactive compounds with potential applications in the food, cosmetic, and therapeutic industries. Among its primary components, fatty acids are particularly noteworth. The oil extracted from the pulp contains approximately 60% oleic acid, 22% palmitic acid, and 12% linoleic acid, along with smaller amounts of palmitoleic and stearic acids. Sterols, including β-sitosterol, stigmasterol, and campesterol, have also been identified. The pulp of açaí is particularly rich in phenolic compounds such as vanillic acid, syringic acid, protocatechuic acid, catechin, and proanthocyanidins, known for their antioxidant properties. These compounds give açaí a high antioxidant capacity, making it a functional food with health benefits, including the neutralization of free radicals and the prevention of chronic diseases [59]. Moreover, the processing of açaí pulp tinto açaí flour maintains the concentration of phenolic compounds. Among the compounds identified are anthocyanins such as cyanidin 3-O-glucoside and cyanidin 3-O-rutinoside (Fig. 6), as well as phenolic acids such a caffeoylquinic acid and flavonoids such as isoorientin, vicenin-2, and isovitexin, which demonstrate potential anti-inflammatory activities. The presence of rhamnocitrin, identified for the first time in açaí by-products, further highlights its antioxidant and anti-inflammatory properties [30].

Açaí seeds, one of the main by-products of the production chain, have been recognized for their various biological activities, which are attributed to their composition of secondary metabolites (Fig. 7). They contain a total phenolic content of 20.63 mg gallic acid equivalents per gram (mg GAE/g), while microwave-assisted extraction (MAE) yielded 18.34 mg GAE/g, and dynamic maceration produced a lower yield of about 8.74 mg GAE/g. The bioaccessibility of the extracted compounds was also

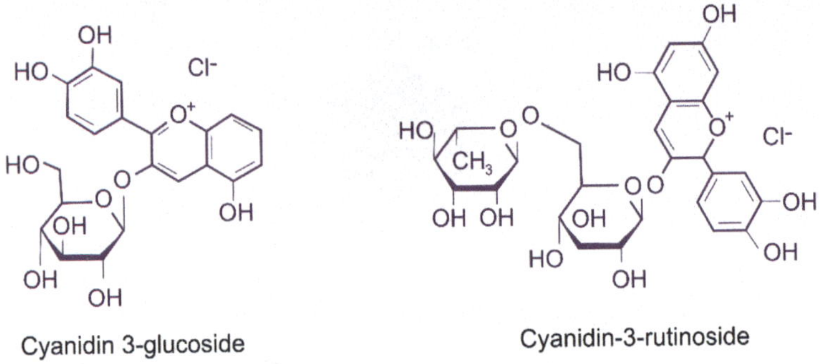

Fig. 6 Chemical structure of the two main anthocyanins found in açaí (*Euterpe oleracea*). *Source* Authors (2024)

evaluated and showed that caftaric acid, quercetin 3-glucoside, and procyanidin B1 were the most bioavailable compounds, with bioaccessibility levels ranging from 36 to 110% in the intestinal phase. These findings suggest that the EDGE extraction process can recover phenolic compounds with potential absorption by the body, underscoring the antioxidant properties of açaí seeds and their applicability in the food and cosmetic industries [20]. Açaí seeds also contain A- and B-type procyanidins, which are oligomers of catechin and epicatechin with an average degree of polymerization of 11.4 (> 3000 Da). For the first time, imaging mass spectrometry (MALDI-IMS) has confirmed the presence of A-type procyanidins in the seed coat (Martins et al. 2020). Additionally, the identification of rhamnocitrin, detected for the first time in açaí by-products, further highlights its antioxidant and anti-inflammatory properties [30].

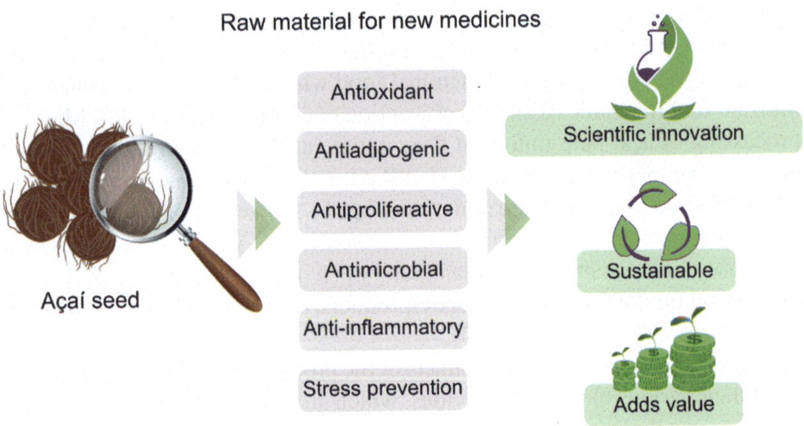

Fig. 7 Some of the proven bioactive properties of the açaí seed: a residue still little explored. *Source* Authors (2024)

Açaí seed extract has demonstrated antioxidant activity by modulating the Nrf2 signaling pathway in human endothelial cells (HUVECs). The extract protected against H_2O_2-induced oxidative stress, preventing cytotoxicity and reactive oxygen species (ROS) production. It also increased the expression of Nrf2-regulated antioxidant enzymes, such as catalase, glutathione peroxidase (GPx), NAD(P)H quinone oxidoreductase 1 (NQO1), and heme oxygenase 1 (HMOX1), suggesting that its protective effect is associated with the activation of this pathway [76]. In studies on mice fed a high-fat diet, the methanolic extract of açaí seeds protected against obesity and adiposity by reducing the adiposity index and triglyceride accumulation in adipocytes. In 3T3-L1 adipocyte cells, açaí inhibited pre-adipocyte proliferation and suppressed adipocyte differentiation, reducing the expression of key proteins involved in adipogenesis, such as PPARγ, FAS, and SREBP-1, suggesting its potential in combating obesity. In studies with mice fed a high-fat diet, methanolic extract of açaí seeds protected against weight gain and adiposity by reducing the adiposity index and the accumulation of triglycerides in adipocytes. In 3T3-L1 adipose cells, açaí inhibited preadipocyte proliferation and suppressed adipocyte differentiation, reducing the expression of key proteins in adipogenesis, such as PPARγ, FAS, and SREBP-1, suggesting its potential to combat obesity [81].

Açaí also has antimicrobial activity, both against gram-positive and gram-negative bacteria, and antifungal activity, making it a potential candidate for the development of antimicrobial products. Hydroalcoholic extracts from different parts of the plant showed significant efficacy against bacteria such as *Clostridium perfringens*, *Staphylococcus aureus* and *Pseudomonas aeruginosa*. In addition, antifungal activity against *Candida albicans* was observed.

Açaí fiber, another by-product of the production chain, is notable for its potential to generate sustainable and biotechnological products, being used in the production of bioethanol and biomethane. Its composition varies seasonally, with a moisture content between 10.5 and 12% and a carbon content between 45 and 52%. With calorific values between 17 and 19 MJ/kg, the fiber is a promising source for energy applications. pretreatment with NaOH facilitates fiber separation and exposes lignin, increasing its suitability for biofuel and biocomposite production [43, 27, 44]. Açaí fibers have significant potential for use in thermoplastic matrix composites due to their good thermal stability and abundant availability [50]. Other studies have also demonstrated the successful reuse of natural açaí fibers in the production of cement-based mortars [53] and soil fertilizers [68, 69].

4.1 *Valorization of Açaí Waste in the Nanobiotechnological Context*

Açaí has been widely used as a raw material in the development of nanobiotechnological products. The use of nanoemulsions and nanoliposomes to stabilize and increase the bioavailability of anthocyanins has been shown to be effective in several experimental evaluations. In addition to improving the release and absorption of anthocyanins in biological models, these formulations provide greater protection against oxidative degradation and, in some cases, increased biological activity [18].

Researchers have demonstrated that the use of açaí oil in nanoemulsion in photodynamic therapy was effective in the treatment of skin cancer (melanoma), in both in vitro and in vivo models. In experiments with B16F10 melanoma cells, the technique resulted in 85% cell death, while maintaining high viability in normal cells. Flow cytometry analysis showed that cell death was predominantly by late apoptosis or necrosis. In tumor-bearing mouse models, treatment with açaí oil-mediated photodynamic therapy reduced tumor volume by 82% compared to the control group. The areas of necrosis in the treated tumor tissue were significantly larger, reinforcing the efficacy of the therapy [55]. The açaí seed oil nanoemulsion showed potent as a therapeutic agent due to its antioxidant properties and selective cytotoxicity against cancer cells. The açaí seed oil and its nanoemulsion showed strong antioxidant activity and antitumor potential against cervical cancer cell lines, such as HeLa and SiHa. The nanoemulsion exhibited a superior antioxidant capacity than the oil alone [11].

Açaí oil-based nanoemulsions have also been developed for the delivery of antifungal drugs, such as ketoconazole, using non-ionic surfactants. Ketoconazole incorporated in the nanoemulsion demonstrated high encapsulation efficiency (98.31%) and good stability. In addition, the açaí nanoemulsions containing ketoconazole presented morphological characteristics suitable for topical use, with spherical and compact droplets. These nanoemulsions have shown promise as a vehicle for the administration of imidazole antifungals, such as ketoconazole, being an effective solution to improve skin absorption, reduce side effects and increase the efficiency of topical treatment of mycoses [21].

In another application, gold nanoparticles synthesized using açaí extracts showed positive effects on tissue repair of palatal wounds in rats. The nanoparticle-treated group showed a reduction in pro-inflammatory markers (TNF-α and IL-1β) compared to the control group, in addition to an increase in anti-inflammatory cytokines (IL-4 and TGF-β). There was also a reduction in oxidative stress levels, with lower concentrations of dichlorofluorescein and nitrite, as well as greater wound contraction and an increase in the collagen area compared to other treatments, suggesting an accelerating effect on the healing process [80].

Another study showed that gold nanoparticles synthesized with açaí extracts showed promising anticancer effects against prostate (PC-3) and pancreatic (Panc-1) cancer cells. Nanoparticles exhibited greater antiproliferative efficacy compared to isolated açaí extracts, especially in pancreatic cancer cells, where cell viability was reduced by up to 41%. Açaí-derived nanoparticles demonstrated selective cytotoxicity, affecting cancer cells without causing significant damage to healthy cells [72].

Due to its bioactive compounds and potential for nanobiotechnological applications, açaí has been used in bioproducts aimed at sustainable development. One example is the açaí oil-based organogel, structured with hyaluronic acid (HA), which demonstrated promising physicochemical properties for cosmetic use. This bioproduct presented a three-dimensional network with effective incorporation of hyaluronic acid, in addition to a spherulitic texture morphology with twisted fibers, indicating a stable crystalline organization. Oxidative stability tests showed that the organogel maintained its organoleptic properties and chemical stability for 90 days under different temperature conditions. Rheological analyses indicated a non-Newtonian viscoelastic behavior, ideal for topical application, facilitating spreadability and viscosity recovery after application [22].

Dry açaí extract has also been used in the field of nutritional supplementation through the development of tablets. When formulated into tablets by direct compression, dry açaí extract showed good mechanical properties and thermal stability. The tablets exhibited weight uniformity, low friability, and adequate mechanical strength, containing 3.65 μg of cyanidin-3-rutinoside per unit. These tablets maintained their physical and chemical properties after exposure to temperatures of up to 200 °C [4].

The use of açaí seeds to produce cellulose and nanocellulose is a sustainable and promising solution that can integrate agro-industrial production chains, reduce waste and promote the development of sustainable materials. Açaí seeds represents up to 90% of the weight of the fruit and are an abundant source of cellulose, with great potential for sustainable production of nanocellulose. The extraction of cellulose from açaí seeds has proven to be viable, taking advantage of a significant portion of the waste generated by the açaí industry, which is typically discarded. The extracted cellulose and nanocellulose have potential applications in packaging, textiles, biomedical devices and environmental remediation, and can replace polymers of petrochemical origin due to their superior mechanical properties and biodegradability [77].

The production of nanocellulose from açaí biomass has properties that vary depending on the extraction method used. Mixed acid hydrolysis is more suitable for applications requiring high thermal stability, while oxidation methods are ideal for applications that requiring good colloidal dispersion and charged surfaces. The nanocellulose extracted by mixed acid hydrolysis exhibited high crystallinity (62%) and thermal stability, but low colloidal stability in water. On the other hand, the nanocelluloses obtained by oxidation presented chemical modification with the conversion of alcoholic groups into carboxylic groups, ensuring a better stability in aqueous suspension [19].

The valorization of the açaí as a fruit and all its residues is emerging as a promising strategy that combines environmental sustainability and technological innovation. The wealth of bioactive compounds presents not only in the pulp, but also in the seeds and fibers of açaí opens a range of opportunities in the food, cosmetics, pharmaceutical and energy industries. The antioxidant, anti-inflammatory, anticancer and antimicrobial properties of these secondary metabolites have been demonstrated in several studies, evidencing their therapeutic and preventive potential in several chronic diseases. Furthermore, nanobiotechnological applications demonstrate how açaí by-products can be transformed into high value-added materials, promoting the circular economy and reducing environmental impact. Thus, açaí stands out not only for its nutritional value, but also as an integral resource whose multiple applications can drive sustainable development and meet the demand for natural and environmentally responsible products.

5 Future Perspectives for Sustainable Açaí Production

The growing global demand for açaí has brought significant economic benefits to the Amazon region, where the açaí palm is native. However, concerns have been raised about the sustainability of its production and the significant waste it generates. Strategies to improve the sustainability of açaí production include innovative agricultural practices and structural changes throughout the açaí value chain. Only by addressing these issues will it be possible to move towards a more sustainable açaí production chain. A key strategy to promote a more sustainable açaí production chain is the implementation of effective public policies [51]. Currently, the açaí market, especially in Brazil, lacks clear and comprehensive regulations to ensure environmentally sustainable production practices [23].

Strengthening environmental regulations and implementing policies that promote sustainable production practices—thereby reducing waste generated throughout the production chain—is essential for protecting biodiversity, curbing deforestation, and encouraging responsible land use across the supply chain, while simultaneously enhancing the overall value of açaí products [23]. In this context, certification schemes that promote sustainable practices have been shown to be a key driver for a more environmentally responsible açaí market [23]. Certifications such as fair trade and organic açaí have been linked to improved environmental management practices by incentivizing producers to adhere to standards that prioritize biodiversity conservation [23, 51]. These certifications also provide access to premium markets where consumers are willing to pay higher prices for sustainably sourced açaí [60]. In addition, certification programs represent an important opportunity for sustainable açaí production, but these programs remain limited in Brazil [60].

In addition, government policies that actively support research and development (R&D) are essential to advancing sustainability in the açaí sector. Historically, the lack of targeted R&D in the industry has limited opportunities for innovation in key areas such as waste valorization, sustainable açaí management, and post-harvest processes [6, 7, 33]. It is necessary to create a network of partnerships between research centers (e.g., public universities and research institutes), governments, and local cooperatives to address these gaps and make significant improvements. Further investment in R&D is essential to ensure that the sector continues to evolve in ways that promote environmental sustainability and economic viability. Santos et al. [67] have shown that substantial investment in R&D is urgently needed to advance the green agenda for forest products, especially non-timber forest products such as açaí. Only by addressing this issue will it be possible to overcome existing knowledge barriers and enable the development of a green bioeconomy [67]. For example, knowledge gaps on the use and applications of açaí waste currently hamper innovation and sustainable use of this forest resource, limiting the potential for bio-based industries to flourish [33, 48]. Therefore, strategic R&D investments would play a key role in addressing these gaps and promoting sustainable practices that support both environmental conservation and economic growth in forest-based sectors.

The açaí production chain has immense potential to become a green economy and drive sustainable development in the Amazon. The açaí industry is closely aligned with several of the United Nations Sustainable Development Goals (SDGs), particularly SDG 12 (responsible consumption and production) and SDG 15 (life on land) [82]. Considering this, the future of açaí production will depend on the integration of sustainable practices, supportive public policies, and technological innovation. By adopting some of the strategies discussed above, the açaí industry can thrive within the framework of a green economy.

6 Conclusions

The açaí fruit and its residues not only hold cultural and economic significance for indigenous and riverine communities but also represent significant economic opportunities for global markets. As the global demand for açaí continues to grow each year, addressing the challenges associated with waste management becomes increasingly urgent, given the substantial amount of waste generated by the açaí production chain. Both the pulp, the most consumed part of the fruit, and its by-products, particularly the seeds, contain a rich diversity of bioactive compounds with antioxidant, anti-inflammatory, and biotechnological properties that warrant exploration.

By valorizing the by-products of the açaí production chain—particularly the seeds and fibers—industries can harness their bioactive potential, fostering innovation and sustainability. This approach not only reduces environmental impacts but also increases the economic value of the açaí supply chain. The strategies outlined in this chapter highlight the importance of integrating waste valorization into sustainable development practices, ultimately contributing to the creation of high-value products with therapeutic applications and broader public health benefits. Through these efforts, the sustainable growth of the açaí industry can be achieved, benefiting both local communities and the global market.

References

1. Albert B, Le Tourneau F (2007) Ethnogeography and resource use among the Yanomami: toward a model of reticular space. Curr Anthropol 48(4):584–592. https://doi.org/10.1086/519914
2. Alessandra-Perini J, Rodrigues-Baptista KC, Machado DE, Nasciutti LE, Perini JA (2018) Anticancer potential, molecular mechanisms and toxicity of *Euterpe oleracea* extract (açaí): a systematic review. PLoS ONE 13(7):e0200101. https://doi.org/10.1371/journal.pone.0200101
3. Anderson AB (1977) Os nomes e usos de palmeiras entre uma tribo de índios Yanomama. Acta Amazon 7(1):5–13. https://doi.org/10.1590/1809-43921977071005
4. Arellano LF, Silva R, Costa DA, Cardoso Da Conceição E, Ribeiro-Costa M, Carrera JO, Junior S (2024) Tablets containing açaí (Euterpe spp.) dried extract for dietary supplementation: a study of functional properties. Food Sci Technol 44:2024. https://doi.org/10.5327/FST.00052

5. Avila-Sosa R, Montero-Rodríguez AF, Aguilar-Alonso P, Vera-López O, Lazcano-Hernández M, Morales-Medina JC, Navarro-Cruz AR (2019) Antioxidant properties of Amazonian fruits: a mini review of in vivo and in vitro studies. Oxid Med Cell Longev 2019(1):8204129. https://doi.org/10.1155/2019/8204129
6. Barbosa JR, de Carvalho Junior RN (2022) Food sustainability trends—how to value the açaí production chain for the development of food inputs from its main bioactive ingredients? Trends Food Sci Technol 124:86–95. https://doi.org/10.1016/J.TIFS.2022.04.005
7. Barbosa PO, Souza MO, Silva MPS, Santos GT, Silva ME, Bermano G, Freitas RN (2021) Açaí (*Euterpe oleracea* Martius) supplementation improves oxidative stress biomarkers in liver tissue of dams fed a high-fat diet and increases antioxidant enzymes' gene expression in offspring. Biomed Pharmacother 139:111627. https://doi.org/10.1016/J.BIOPHA.2021.111627
8. Barros L, Calhelha RC, Queiroz MJRP, Santos-Buelga C, Santos EA, Regis WCB, Ferreira ICFR (2015) The powerful in vitro bioactivity of *Euterpe oleracea* Mart. seeds and related phenolic compounds. Ind Crops Prod 76:318–322. https://doi.org/10.1016/J.INDCROP.2015.05.086
9. Bellucci ERB, dos Santos JM, Carvalho LT, Borgonovi TF, Lorenzo JM, da Silva-Barretto AC (2022) Açaí extract powder as natural antioxidant on pork patties during the refrigerated storage. Meat Sci 184:108667. https://doi.org/10.1016/J.MEATSCI.2021.108667
10. Borges PRS, Tavares EG, Guimarães IC, Rocha RDP, Araujo ABS, Nunes EE, Boas EVDBV (2016) Obtaining a protocol for extraction of phenolics from açaí fruit pulp through Plackett-Burman design and response surface methodology. Food Chem 210:189–199. https://doi.org/10.1016/J.FOODCHEM.2016.04.077
11. Borges KRA, Wolff LAS, da Silva MACN, de Carvalho Silva AK, Campos CDL, Souza FS, Teles AM, Vale AÁM, Pascoa H, Lima EM, de Sousa EM, Nunes ACS, Gil da Costa RM, Faustino-Rocha AI, Cardoso Carvalho R, Nascimento M do DSB (2024) Açaí (*Euterpe oleracea* Mart.) seed oil and its nanoemulsion: chemical characterisation, toxicity evaluation, antioxidant and anticancer activities. Curr Issues Mol Biol 46(5):3763–3793. https://doi.org/10.3390/CIMB46050235
12. Bufalino L, Guimarães AA, Silva BMDSE, De Souza RLF, De Melo ICNA, De Oliveira DNPS, Trugilho PF (2018) Local variability of yield and physical properties of açaí waste and improvement of its energetic attributes by separation of lignocellulosic fibers and seeds. J Renew Sustain Energy 10(5). https://doi.org/10.1063/1.5027232
13. Buratto RT, Cocero MJ, Martín Á (2021) Characterization of industrial açaí pulp residues and valorization by microwave-assisted extraction. Chem Eng Process Process Intensification 160:108269. https://doi.org/10.1016/J.CEP.2020.108269
14. Buratto RT, Hoyos EG, Cocero MJ, Martín Á (2019) Impregnation of açaí residue extracts in silica-aerogel. J Supercrit Fluids 146:120–127. https://doi.org/10.1016/J.SUPFLU.2018.12.004
15. Calzavara BBG (1972) As possibilidades do açaizeiro no estuário amazônico
16. Cavalcante PB (1974) Frutas comestíveis da Amazônia. Museu Paraense Emílio Goeldi, Belém
17. Cedrim PCAS, Barros EMA, do Nascimento TG (2018) Propriedades antioxidantes do açaí (*Euterpe oleracea*) na síndrome metabólica. Braz J Food Technol 21:e2017092. https://doi.org/10.1590/1981-6723.09217
18. Chen BH, Inbaraj BS (2019) Nanoemulsion and nanoliposome based strategies for improving anthocyanin stability and bioavailability. Nutrients 11(5). https://doi.org/10.3390/NU11051052
19. Cidreira ACM, Hatami T, Linan LZ, Pinheiro IF, Gomes RC, Rocha JJC, Mei LHI (2024) Nanocellulose extraction from acai bagasse through mixed acid hydrolysis and oxidative techniques. Int J Biol Macromol 273(Pt 1). https://doi.org/10.1016/J.IJBIOMAC.2024.133034
20. Conrado NM, dos Santos PNA, da Conceição Prudêncio Dutra M, Krause LC, dos Santos Polidoro A, dos Santos Lima M, dos Santos AL, Caramão EB (2024) Energized dispersive guided extraction (EDGE), a new extractive approach of phenolics from Açaí (E. oleracea) seeds: chemical characterization, antioxidant properties, and bioaccessibility of the extracts. Food Bioproc Tech 1–13. https://doi.org/10.1007/S11947-024-03490-1/FIGURES/2

21. Contente DML, Pereira RR, Rodrigues AMC, da Silva EO, Ribeiro-Costa RM, Carrera Silva-Júnior JO (2020) Nanoemulsions of Acai oil: physicochemical characterization for the topical delivery of antifungal drugs. Chem Eng Technol 43(7):1424–1432. https://doi.org/10.1002/CEAT.201900627
22. da Sanches SC, Ré MI, Silva-Júnior JOC, Ribeiro-Costa RM (2023) Organogel of acai oil in cosmetics: microstructure, stability, rheology and mechanical properties. Gels 9(2). https://doi.org/10.3390/GELS9020150
23. Damasco G, Anhalt M, Perdiz RO, Wittmann F, de Assis RL, Schöngart J, Piedade MTF, Bacon CD, Antonelli A, Fine PVA (2022) Certification of açaí agroforestry increases the conservation potential of the Amazonian tree flora. Agrofor Syst 96(2):407–416. https://doi.org/10.1007/S10457-021-00727-2/FIGURES/3
24. de Sousa FF, Vieira-da-Silva C, Barros FB (2018) The (in)visible market of miriti (*Mauritia flexuosa* L.f.) fruits, the "winter acai", in Amazonian riverine communities of Abaetetuba, Northern Brazil. Glob Ecol Conserv 14. https://doi.org/10.1016/J.GECCO.2018.E00393
25. de Vilhena Amanajás Miranda L, Mochiutti S, da Cunha AC, Cunha HFA (2022) Discarding and final destination of açaí in the Oriental Amazon—Brazil. Ambiente Soc 25. https://doi.org/10.1590/1809-4422asoc20200138r2vu2022l2oa
26. de Mesquita AL, Barrero NG, Fiorelli J, Christoforo AL, De Faria LJG, Lahr FAR (2018) Eco-particleboard manufactured from chemically treated fibrous vascular tissue of acai (*Euterpe oleracea* Mart.) fruit: a new alternative for the particleboard industry with its potential application in civil construction and furniture. Ind Crops Prod 112:644–651. https://doi.org/10.1016/J.INDCROP.2017.12.074
27. de Oliveira DNPS, Claro PIC, de Freitas RR, Martins MA, Souza TM, da Silva BM, Mendes LM, Bufalino L (2019) Enhancement of the Amazonian açaí waste fibers through variations of alkali pretreatment parameters. Chem Biodivers 16(9). https://doi.org/10.1002/cbdv.201900275
28. de Oliveira M do SP, de Farias Neto JT, de Mattieto RA, Mochiutti S, Carvalho AV (2017) Açaí: *Euterpe oleracea*. Instituto Interamericano de Cooperación para la Agricultura (IICA), Buenos Aires
29. de Oliveira M do SP, Carvalho JEU, Nascimento WMO (2000) Açaí *Euterpe oleracea* Mart. Funep, Jaboticabal
30. de Silva APS, de Camargo AC, Lazarini JG, Franchin M, de Sardi JCO, Rosalen PL, de Alencar SM (2022) Phenolic profile and the antioxidant, anti-inflammatory, and antimicrobial properties of açaí (*Euterpe oleracea*) meal: a prospective study. Foods 12(1):86. https://doi.org/10.3390/FOODS12010086
31. Do Nascimento VHN, Lima CDS, Paixão JTC, Freitas JJDS, Kietzer KS (2016) Antioxidant effects of açaí seed *Euterpe oleracea* in anorexia-cachexia syndrome induced by Walker-256 tumor. Acta Cir Bras 31(9):597–601. https://doi.org/10.1590/S0102-865020160090000004
32. Farinas CS, dos Santos RRM, Bertucci Neto V, Pessoa JD (2009) Aproveitamento do Caroço do Açaí como Substrato para a Produção de Enzimas por Fermentação em Estado Sólido. Embrapa, São Carlos, São Paulo
33. Ferreira SF, Buller LS, Maciel-Silva FW, Sganzerla WG, Berni MD, Forster-Carneiro T (2021) Waste management and bioenergy recovery from açaí processing in the Brazilian Amazonian region: a perspective for a circular economy. Biofuels Bioprod Biorefin 15(1):37–46. https://doi.org/10.1002/BBB.2147
34. Georgescu-Roegen N (1975) Bio-economic aspects of entropy, entropy and information in science and philosophy
35. Hogan S, Chung H, Zhang L, Li J, Lee Y, Dai Y, Zhou K (2010) Antiproliferative and antioxidant properties of anthocyanin-rich extract from açai. Food Chem 118(2):208–214. https://doi.org/10.1016/J.FOODCHEM.2009.04.099
36. IBGE (2024) Produção de Açaí no Brasil. https://www.ibge.gov.br/explica/producao-agropecuaria/acai-cultivo/br. Accessed 28 Oct 2024
37. IBGE IB de G e E (2024) Produção Agrícola Municipal

38. Igreja WS, da Silva Martins LH, de Almeida RR, de Oliveira JAR, Lopes AS, Chisté RC (2023) Açai seeds (*Euterpe oleracea* Mart) are agroindustrial waste with high potential to produce low-cost substrates after acid hydrolysis. Molecules 28(18):6661. https://doi.org/10.3390/MOLECULES28186661/S1
39. Instituto Brasileiro de Geografia e Estatística I (2018) Censo Agro 2017. In: Censo Agropecuário. https://censoagro2017.ibge.gov.br/resultados-censo-agro-2017.html. Accessed 6 Oct 2024
40. Koide RT, Nguyen BT, Skinner RH, Dell CJ, Peoples MS, Adler PR, Drohan PJ (2015) Biochar amendment of soil improves resilience to climate change. GCB Bioenergy 7(5):1084–1091. https://doi.org/10.1111/GCBB.12191
41. Laurindo LF, Barbalho SM, Araújo AC, Guiguer EL, Mondal A, Bachtel G, Bishayee A (2023) Açaí (*Euterpe oleracea* Mart.) in health and disease: a critical review. Nutrients 15(4). https://doi.org/10.3390/NU15040989
42. Lay CZE, da Cruz J, Blair-Matos C, Andrade Luzeiro JA, Souza Matos JC, de Oliveira MC, Bernardes Filho LA, Pereira Barbosa A (2021) Traditional knowledge of açaí (*Euterpe precatoria* Mart.—Arecaceae) usage in the sustainable development reserve—RDS Piagaçu Purus—Amazonas—Brazil. Int J Environ Agric Biotechnol 6(1):127–134. https://doi.org/10.22161/IJEAB.61.17
43. de Lima ACP, Bastos DLR, Camarena MA, Bon EPS, Cammarota MC, Teixeira RSS, Gutarra MLE (2021) Physicochemical characterization of residual biomass (seed and fiber) from açaí (*Euterpe oleracea*) processing and assessment of the potential for energy production and bioproducts. Biomass Convers Biorefin 11(3):925–935. https://doi.org/10.1007/S13399-019-00551-W/FIGURES/6
44. de Lima ACP, Bastos DLR, Camarena MA, Bon EPS, Cammarota MC, Teixeira RSS, Gutarra MLE (2021) Physicochemical characterization of residual biomass (seed and fiber) from açaí (*Euterpe oleracea*) processing and assessment of the potential for energy production and bioproducts. Biomass Convers Biorefin 11(3):925–935. https://doi.org/10.1007/s13399-019-00551-w
45. Lucas BF, Zambiazi RC, Costa JAV (2018) Biocompounds and physical properties of açaí pulp dried by different methods. LWT 98:335–340. https://doi.org/10.1016/J.LWT.2018.08.058
46. MAPA, Ministério da Agricultura P e A (2023) Registro Nacional de Cultivares. In: CultivarWeb
47. Machado AK, Cadoná FC, Assmann CE, Andreazza AC, Duarte MMMF, dos Santos BC, Zhou X, de Souza DV, Ribeiro EE, da Cruz IBM (2019) Açaí (*Euterpe oleracea* Mart.) has anti-inflammatory potential through NLRP3-inflammasome modulation. J Funct Foods 56:364–371. https://doi.org/10.1016/J.JFF.2019.03.034
48. Macêdo N, Macêdo M, Araujo P, Berni P, Zandoná LR, Zandoná Z, Maria N, De Toledo V, Da Silva PPM, Ang´ A, De Toledo AA, Ario M´, Maróstica R, Junior MM (2024) Potential of Brazilian berries in developing innovative, healthy, and sustainable food products. Sustain Food Technol 2(3):506–530. https://doi.org/10.1039/D3FB00130J
49. Market Date Forecast (2024) Acai Berry Market size, share, growth report | 2024 to 2032. https://www.marketdataforecast.com/market-reports/acai-berry-market. Accessed 28 Oct 2024
50. Martins MA, Pessoa JDC, Gonçalves PS, Souza FI, Mattoso LHC (2008) Thermal and mechanical properties of the açaí fiber/natural rubber composites. J Mater Sci 43(19):6531–6538. https://doi.org/10.1007/S10853-008-2842-4/FIGURES/7
51. Martins CM, Pereira SCF, Scarpin MRS, Queiroz MM, da Silva Cavalcante M (2024) The impact of customer pressures and government regulations on the implementation of socio-environmental practices in organic certification in the Amazon region. Benchmarking ahead-of-p (ahead-of-print). https://doi.org/10.1108/BIJ-07-2023-0453/FULL/PDF
52. Martins GR, do Amaral FRL, Brum FL, Mohana-Borges R, de Moura SS, Ferreira FA, Sangenito LS, Santos AL, Figueiredo NG, da Silva ASA (2020) Chemical characterization, antioxidant and antimicrobial activities of açaí seed (Euterpe oleracea Mart.) extracts containing A- and B-type procyanidins. LWT. https://doi.org/10.1016/j.lwt.2020.109830

53. Marvila MT, Azevedo ARG, Cecchin D, Costa JM, Xavier GC, de Fátima do Carmo D, Monteiro SN (2020) Durability of coating mortars containing açaí fibers. Case Stud Constr Mater 13:e00406. https://doi.org/10.1016/J.CSCM.2020.E00406
54. Melo PS, Selani MM, Gonçalves RH, de Paulino JO, Massarioli AP, de Alencar SM (2021) Açaí seeds: an unexplored agro-industrial residue as a potential source of lipids, fibers, and antioxidant phenolic compounds. Ind Crops Prod 161:113204. https://doi.org/10.1016/J.INDCROP.2020.113204
55. Monge-Fuentes V, Muehlmann LA, Longo JPF, Silva JR, Fascineli ML, Azevedo RB, de Souza P, Faria F, Degterev IA, Rodriguez A, Carneiro FP, Lucci CM, Escobar P, Amorim RFB (2017) Photodynamic therapy mediated by acai oil (*Euterpe oleracea* Martius) in nanoemulsion: a potential treatment for melanoma. J Photochem Photobiol B 166:301–310. https://doi.org/10.1016/J.JPHOTOBIOL.2016.12.002
56. Neri-Numa IA, Soriano Sancho RA, Pereira APA, Pastore GM (2018) Small Brazilian wild fruits: nutrients, bioactive compounds, health-promotion properties and commercial interest. Food Res Int 103:345–360. https://doi.org/10.1016/J.FOODRES.2017.10.053
57. Neves EG, Heckenberger MJ (2019) The call of the wild: rethinking food production in ancient Amazonia. Annu Rev Anthropol 48(1):371–388. https://doi.org/10.1146/annurev-anthro-102218-011057
58. Nogueira OL, de Carvalho CJR, Muller CH, Galvão EUP, Silva HM e, Rodrigues JELF, Oliveira M do SP, Carvalho JEU, da Rocha Neto OG, do Nascimento WMO, Calzavara BBG (1995) A cultura do açaí. Embrapa: Centro de Pesquisa Agroflorestal da Amazônia Oriental, Brasília
59. Pacheco-Palencia LA, Mertens-Talcott S, Talcott ST (2008) Chemical composition, antioxidant properties, and thermal stability of a phytochemical enriched oil from acai (*Euterpe oleracea* Mart.). J Agric Food Chem 56(12):4631–4636. https://doi.org/10.1021/JF800161U
60. Pepper LG, De Freitas L, Alves N, Alves LDFN (2024) Small-scale açaí in the global market: adding value to ensure sustained income for forest farmers in the Amazon estuary 389–413. https://doi.org/10.1007/978-3-031-54270-1_12
61. Pessoa JDC, Arduin M, Martins MA, de Carvalho JEU (2010) Characterization of açaí (*E. oleracea*) fruits and its processing residues. Braz Arch Biol Technol 53(6):1451–1460. https://doi.org/10.1590/S1516-89132010000600022
62. Pessôa TS, de Lima Ferreira LE, da Silva MP, Neto LMP, do Nascimento BF, Fraga TJM, Jaguaribe EF, Cavalcanti JV, da Motta Sobrinho MA (2019) Açaí waste beneficing by gasification process and its employment in the treatment of synthetic and raw textile wastewater. J Clean Prod 240:118047. https://doi.org/10.1016/J.JCLEPRO.2019.118047
63. Pratt SG, Matthews K (2006) Superfoods: fourteen foods that will change your life. HarperCollins Publishers
64. Quaresma AP, Euler AMC (2023) Açaí, mais que um fruto, símbolo da cultura alimentar e bioeconomia da Amazônia. In: Bioeconomia e o mercado dos produtos florestais não madeireiros: desafios e possibilidades. Synergia Consultoria, São Paulo, pp 74–99
65. Queiroz JAL, Mochiutti S (2012) Guia prático de manejo de açaizais para produção de frutos
66. Saha A, Basak BB (2020) Scope of value addition and utilization of residual biomass from medicinal and aromatic plants. Ind Crops Prod 145:111979. https://doi.org/10.1016/J.INDCROP.2019.111979
67. Santos JPB, Romanelli JP, Gardon FR, Krainovic PM, de Resende AF, Souza LR, Piotto D, Rodrigues RR (2023) Multifunctional forest restoration in Brazil: a critical analysis of the trends and knowledge gaps in the scientific literature. Sustainability 15(22):15782. https://doi.org/10.3390/SU152215782
68. Sato MK, de Lima HV, Costa AN, Rodrigues S, Pedroso AJS, de Freitas Maia CMB (2019) Biochar from acai agroindustry waste: study of pyrolysis conditions. Waste Manage 96:158–167. https://doi.org/10.1016/J.WASMAN.2019.07.022
69. Sato MK, de Lima HV, Noronha Costa A, Rodrigues S, Mooney SJ, Clarke M, Silva Pedroso AJ, de Freitas Maia CMB (2020) Biochar as a sustainable alternative to açaí waste disposal in Amazon, Brazil. Process Saf Environ Prot 139:36–46. https://doi.org/10.1016/J.PSEP.2020.04.001

70. Schroth G, Da Mota MDSS (2014) Agroforestry: complex multistrata agriculture. In: Encyclopedia of agriculture and food systems. Elsevier, pp 195–207
71. Sganzerla WG, Ampese LC, Parisoto TAC, Forster-Carneiro T (2023) Process intensification for the recovery of methane-rich biogas from dry anaerobic digestion of açaí seeds. Biomass Convers Biorefin 13(9):8101–8114. https://doi.org/10.1007/S13399-021-01698-1/FIGURES/5
72. Sibuyi NRS, Thipe VC, Panjtan-Amiri K, Meyer M, Katti KV (2021) Green synthesis of gold nanoparticles using Acai berry and Elderberry extracts and investigation of their effect on prostate and pancreatic cancer cells. Nanobiomedicine (Rij) 8:1–8. https://doi.org/10.1177/1849543521995310
73. da Silva TR, de Matos PR, Tambara Júnior LUD, Marvila MT, de Azevedo ARG (2023) A review on the performance of açaí fiber in cementitious composites: characteristics and application challenges. J Build Eng 71:106481. https://doi.org/10.1016/J.JOBE.2023.106481
74. Silva TOM (2017) Compostos voláteis, perfil de aroma, de ácidos graxos e potencial antioxidante de óleo extraído por prensagem a frio de resíduo agroindustrial de açaí (*Euterpe oleracea* mart.). Universidade Federal de Sergipe
75. da Silveira JT, da Rosa APC, de Morais MG, Victoria FN, Costa JAV (2023) An integrative review of açaí (*Euterpe oleracea* and *Euterpe precatoria*): traditional uses, phytochemical composition, market trends, and emerging applications. Food Res Int 173:113304. https://doi.org/10.1016/J.FOODRES.2023.113304
76. Soares ER, Monteiro EB, de Bem GF, Inada KOP, Torres AG, Perrone D, Soulage CO, Monteiro MC, Resende AC, Moura-Nunes N, Costa CA, Daleprane JB (2017) Up-regulation of Nrf2-antioxidant signaling by açaí (*Euterpe oleracea* Mart.) extract prevents oxidative stress in human endothelial cells. J Funct Foods 37:107–115. https://doi.org/10.1016/J.JFF.2017.07.035
77. Teixeira LHC, de Morais MG, Costa JAV (2024) Sustainable cellulose and nanocellulose production from Amazon forest açaí residues. Biomass Convers Biorefin 1–9. https://doi.org/10.1007/S13399-024-06073-4/FIGURES/2
78. Teodoro AJ (2019) Bioactive compounds of food: their role in the prevention and treatment of diseases. Oxid Med Cell Longev 2019. https://doi.org/10.1155/2019/3765986
79. Ter Steege H, Pitman NCA, Sabatier D, Baraloto C, Salomão RP, Guevara JE, Phillips OL, Castilho CV, Magnusson WE, Molino J-F, Monteagudo A, Núñez Vargas P, Montero JC, Feldpausch TR, Coronado ENH, Killeen TJ, Mostacedo B, Vasquez R, Assis RL, Terborgh J, Wittmann F, Andrade A, Laurance WF, Laurance SGW, Marimon BS, Marimon B-H, Guimarães Vieira IC, Amaral IL, Brienen R, Castellanos H, Cárdenas López D, Duivenvoorden JF, Mogollón HF, Matos FDDA, Dávila N, García-Villacorta R, Stevenson Diaz PR, Costa F, Emilio T, Levis C, Schietti J, Souza P, Alonso A, Dallmeier F, Montoya AJD, Fernandez Piedade MT, Araujo-Murakami A, Arroyo L, Gribel R, Fine PVA, Peres CA, Toledo M, Aymard C. GA, Baker TR, Cerón C, Engel J, Henkel TW, Maas P, Petronelli P, Stropp J, Zartman CE, Daly D, Neill D, Silveira M, Paredes MR, Chave J, Lima Filho DDA, Jørgensen PM, Fuentes A, Schöngart J, Cornejo Valverde F, Di Fiore A, Jimenez EM, Peñuela Mora MC, Phillips JF, Rivas G, Van Andel TR, Von Hildebrand P, Hoffman B, Zent EL, Malhi Y, Prieto A, Rudas A, Ruschell AR, Silva N, Vos V, Zent S, Oliveira AA, Schutz AC, Gonzales T, Trindade Nascimento M, Ramirez-Angulo H, Sierra R, Tirado M, Umaña Medina MN, Van Der Heijden G, Vela CIA, Vilanova Torre E, Vriesendorp C, Wang O, Young KR, Baider C, Balslev H, Ferreira C, Mesones I, Torres-Lezama A, Urrego Giraldo LE, Zagt R, Alexiades MN, Hernandez L, Huamantupa-Chuquimaco I, Milliken W, Palacios Cuenca W, Pauletto D, Valderrama Sandoval E, Valenzuela Gamarra L, Dexter KG, Feeley K, Lopez-Gonzalez G, Silman MR (2013) Hyperdominance in the {Amazonian} {Tree} {Flora}. Science 342(6156):1243092 (1979). https://doi.org/10.1126/science.1243092
80. Thirupathi A, Guzzatti MFM, Corrêa MEAB, Venturini LM, de Casagrande LR, Lima IR, Da Costa C, De Pieri E, Tietbohl LTW, Feuser PE, Machado-de-Ávila RA, Gu Y, Silveira PCL (2023) Green synthesis of gold nanoparticles with curcumin or açai in the tissue repair of palatal wounds. Antioxidants 12(8):1574. https://doi.org/10.3390/ANTIOX12081574/S1

81. Trindade PL, dos Soares ER, Monteiro EB, Resende ÂC, Moura-Nunes N, Souza-Mello V, Ferraz DC, Daleprane JB (2019) Antiadipogenic effects of açai seed extract on high fat diet-fed mice and 3T3-L1 adipocytes: a potential mechanism of action. Life Sci 228:316–322. https://doi.org/10.1016/J.LFS.2019.04.051
82. UN (2020) THE 17 GOALS | Sustainable Development
83. Vianna SA (2020a) *Euterpe oleracea* in Flora do Brasil
84. Vianna SA (2020) Euterpe precatoria in Flora do Brasil. https://floradobrasil.jbrj.gov.br/consulta/?grupo=6&familia=null&genero=Euterpe&especie=precatoria&autor=&nomeVernaculo=&nomeCompleto=&formaVida=null&substrato=null&ocorreBrasil=QUALQUER&ocorrencia=OCORRE&endemismo=TODOS&origem=TODOS&regiao=QUALQUER&ilhaO. Accessed 4 Oct 2024
85. Viganó J, de Aguiar AC, Veggi PC, Sanches VL, Rostagno MA, Martínez J (2022) Techno-economic evaluation for recovering phenolic compounds from acai (*Euterpe oleracea*) by-product by pressurized liquid extraction. J Supercrit Fluids 179:105413. https://doi.org/10.1016/J.SUPFLU.2021.105413
86. Yamaguchi KKDL, Pereira LFR, Lamarão CV, Lima ES, Da Veiga-Junior VF (2015) Amazon acai: chemistry and biological activities: a review. Food Chem 179:137–151. https://doi.org/10.1016/J.FOODCHEM.2015.01.055

Challenges, Opportunities, and Future Perspectives Regarding the Valorization of Agro-industrial Waste

Alfredo José dos Santos Junior, Álison Moreira da Silva, Kamilla Crysllayne Alves da Silva, Allana Katiussya Silva Pereira, and Ananias Francisco Dias Júnior

Abstract The concept of agroindustry is an environment that encompasses activities related to the three classical sectors of the economy: primary (agriculture, livestock, forestry, and aquaculture), secondary (transformation and processing), and tertiary (distribution). The agroindustry is critical in supplying food and essential materials for modern life. However, significant quantities of products derived from the agro-industrial sector generate significant quantities of waste. The utilization of agro-industrial waste becomes a challenging process because the material disposed of exhibits considerable heterogeneity in its composition, hindering the adoption of large-scale measures to reduce potential damage. To utilize agro-industrial waste, one must overcome numerous challenges, beginning with socioeconomic barriers and extending to technological limitations related to equipment or waste conversion methods. One alternative to overcoming the socioeconomic barrier is employing agro-industrial waste converted into high-value bioproducts in various commercial sectors, such as pharmaceuticals, food, energy generation, and chemical industries. As for the technological challenges of agro-industrial waste utilization, there are several ongoing research initiatives, innovation efforts, and regulations aimed at improving the utilization of this material in various sectors, with particular emphasis

A. J. dos Santos Junior (✉) · K. C. A. da Silva · A. K. S. Pereira
Department of Forest Sciences, "Luiz de Queiroz" College of Agriculture (Esalq), Universidade de São Paulo (USP), Piracicaba, São Paulo, Brazil
e-mail: alf.junior12@gmail.com

K. C. A. da Silva
e-mail: kamilla.alves@usp.br

A. K. S. Pereira
e-mail: allana.florestal@gmail.com

Á. M. da Silva · A. F. Dias Júnior
Agricultural Sciences and Engineering Center, Universidade Federal Do Espírito Santo (Ufes), Jerônimo Monteiro, Espírito Santo, Brazil
e-mail: alison_vni@hotmail.com

A. F. Dias Júnior
e-mail: ananias.dias@ufes.br

A. K. Silva Pereira and A. F. Dias Júnior (eds.), *Agro-industrial Waste for New Pharmaceuticals*, Green Energy and Technology,
https://doi.org/10.1007/978-3-031-88534-1_5

on the food and pharmaceutical industries. A growing trend in agro-industrial waste utilization involves the concept of biorefineries, where processed organic material with minimal or no waste is entirely transformed into high-value products suitable for commercialization or energy generation. This chapter will address the main challenges and potential solutions for using agro-industrial waste to reduce potential environmental harm and add value to materials that would otherwise be discarded.

Keywords Socio-economic aspects of agro-industrial waste · Technical challenges · Technological innovation · Bioproducts · Waste valorization

1 Introduction

Have you ever imagined a world where every waste generated by the agroindustry is transformed into valuable resources, driving economies and helping to combat climate change? This world is yet to be the reality, and we need to move toward a tipping point where what was once considered raw material waste becomes a new source of innovation. This scenario raises a crucial question: are we indeed prepared to valorize, on a global scale, the billions of tons of agro-industrial waste we generate annually?

The agroindustry is fundamental in providing food and essential materials for modern life. However, its growth comes with a significant environmental cost. In addition to food waste, estimated by the FAO at 1.05 billion tons in 2022 [93], non-edible residues such as peels, pomace, and straw present a growing challenge. According to Shinde et al. [85], global agricultural waste production reaches about 5 billion tons annually. Asia is the largest producer, contributing 47% of this total, followed by the Americas with 29%, Europe with 16%, Africa with 7%, and Oceania with 1%. When not correctly reused, these wastes can potentially cause serious environmental problems, contributing to soil degradation, contamination of water resources, and greenhouse gas emissions such as methane and carbon dioxide [12, 87]. However, the contemporaneous vision of a circular economy invites us to reimagine the role of these wastes. The central concept is transforming agricultural and industrial by-products into valuable inputs for other sectors, creating a more efficient and sustainable production chain. The "zero waste" idea is gaining momentum, with the reutilization of agro-industrial waste at the heart of this transformation. However, what exactly does this shift entail?

The technical challenges are complex. The variability in chemical composition and physical properties demands the development of specific technologies for each type of waste [7, 12]. Furthermore, transporting these materials, often dispersed in rural areas, presents a logistical and financial obstacle [7]. It is necessary to think of local solutions that are economically viable and environmentally sustainable. In this regard, biorefineries emerge as one of the most promising solutions [82, 101]. They operate similarly to oil refineries but utilize biomass, such as agro-industrial waste, to produce a wide range of products: biofuels, biochemicals, bioplastics, and fertilizers [12, 82, 101].

These multifunctional facilities are critical to efficient waste valorization, enabling the extraction of various high-value products from a single raw material source. Biorefinery operations generate minimal waste, thus reducing pressure on the environment and ecosystem services [82], representing a significant leap in efficiency and sustainability. However, the path to the large-scale implementation of these technologies is still long. The initial costs of setting up biorefineries and the need for continuous research and development represent substantial barriers [82, 89]. While promising, many of these technologies are still in the development phase or are economically unfeasible on a large scale, especially in developing countries with limited infrastructure. How can these barriers be overcome?

At this point, interdisciplinary collaboration comes into play. Researchers, governments, industry, and society must work together to create an environment conducive to developing these innovations. Research institutions play a crucial role in driving technological advancement, and governments must create public policies encouraging this progress, such as subsidies for clean technologies and regulations promoting the circular economy. The private sector, in turn, must view waste not as a burden but as a business opportunity. Finally, society must know the benefits of sustainable practices and demand change.

The circular economy has become a priority for many countries aiming to align economic growth with sustainability. In Europe, initiatives such as the European Union's Circular Economy Action Plan promote waste reuse, recycling, and reduction across various production chains [33]. This plan also includes specific targets for electronics, plastics, textiles, and construction sectors, which could drive regional gross domestic product (GDP) growth and create new jobs by 2030. In Latin America, several countries have adopted measures focused on the circular economy. Brazil, for example, created the National Circular Economy Strategy (ENEC) [18], which aims to optimize resource use and extend product lifespans by encouraging reuse and recycling. A proposed law (PL 1874/2022) is under consideration, setting targets to reduce waste, promote producer responsibility, and encourage research into sustainable technologies [81]. These policies reinforce the concept that a circular economy can positively impact both the environment and long-term economic development, with many countries working to strengthen their regulations to enable this sustainable transition.

The economic impact of an approach focused on the valorization of agro-industrial waste is immense. According to estimates from the United Nations Environment Programme [94], the circular economy could generate $108 billion for the global economy by 2050, including new market opportunities in bioplastics, bioenergy, and biofertilizers. Moreover, these practices could generate millions of jobs, particularly in rural regions and developing areas, where waste reutilization could strengthen the local economy and provide new sources of income for small farmers and rural workers.

However, the economic challenges go beyond the simple adoption of technologies. Creating an efficient market for products derived from agro-industrial waste requires constant demand and a well-structured supply chain. Many of the products generated from waste still face resistance, whether due to consumer lack of knowledge, cultural issues, or an inefficient strategic approach by marketing in promoting sustainability-focused products [37, 74, 76]. For example, bioplastics made from agricultural waste still need to compete significantly with conventional plastics in terms of price despite their environmental advantages [76]. To overcome this obstacle, public policies such as tax incentives, eco-labeling, and awareness campaigns can play a crucial role in creating broader demand for these products.

In addition to the economic aspects, there are also significant social considerations. Integrating agro-industrial waste valorization technologies in small and medium-sized agro-industrial enterprises can be a powerful lever for social development, particularly in countries with large rural populations [34, 73]. Adopting these practices can improve the living conditions of farming communities by providing new ways to add value to products and increase household income. Training and education programs will also be essential to prepare these communities for emerging opportunities, ensuring everyone can benefit from the transition to a more circular and sustainable economy.

We must recognize the innovation opportunities that emerge when we combine traditional agricultural knowledge with new technologies. In various regions worldwide, local communities are already efficiently utilizing agro-industrial waste for energy generation or composting [39, 61]. The combination of this ancestral knowledge with modern science can lead to even more innovative and sustainable solutions. Another relevant point is the growing research in biotechnology applied to waste valorization. Researchers are developing specific enzymes to break down lignocellulose present in wastes such as straw and rice husks, enabling the production of second-generation biofuels [29]. These advancements can reduce dependence on fossil fuels while providing a sustainable outlet for waste that would otherwise be discarded or incinerated [29]. Additionally, advanced fermentation techniques are being developed to convert agricultural waste into high-value products, such as bioactive compounds, which can be used in pharmaceutical and cosmetic industries [27, 31].

Ultimately, this chapter proposes a profound reflection on the challenges and opportunities related to the valorization of agro-industrial waste. The transition to a circular economy is an environmental necessity and an opportunity to create an economic and social paradigm where waste becomes a resource, waste transforms into opportunity, and sustainability drives growth. As climate and environmental crises intensify, the valorization of agro-industrial waste offers a real alternative to mitigate environmental damage and promote sustainable development. However, are we willing to face the necessary challenges to seize this opportunity as a society? The answer to this question will define the future of the coming generations.

2 Socio-economic Aspects of Agro-industrial Waste Valorization

The transition to a green economy, where agro-industrial waste is transformed into valuable resources, goes far beyond a simple technological advancement. It represents a socioeconomic revolution with the potential to reshape entire industries, create jobs, reduce inequalities, and promote more sustainable and inclusive development. In a world where population growth and resource demand increasingly pressure productive systems, the question is not only what can be done with agro-industrial waste but who can benefit from this transformation and how these benefits will be distributed. In this context, waste can become more than an environmental problem; it can become a new path toward economic, social, and environmental justice.

According to the United Nations Environment Programme (UNEP), about one-third of the food produced globally is lost or wasted throughout the supply chain, resulting in approximately 1.05 billion tons of waste in 2022 [93]. Much of this waste occurs in the agro-industrial sector, which encompasses everything from harvesting to food processing (Fig. 1). Instead of being discarded, these wastes present an unprecedented opportunity to transform economies, particularly in regions where agriculture is central to socioeconomic development.

The valorization of agro-industrial waste is not merely a matter of efficiency. It represents a profound shift in how we perceive economic cycles. Instead of linear models based on resource extraction, production, and disposal, we are moving towards circular models, where waste becomes an input for new production processes [34, 82]. This shift directly impacts job creation and income distribution, as it enables the formation of new value chains, many of which can be accessed by small and medium-sized farmers, cooperatives, and local entrepreneurs.

An emblematic example is the use of agro-industrial waste in biofuel production. With its vast sugarcane production, Brazil has long utilized sugarcane bagasse as an energy source. In 2022, Brazilian sugar-energy mills contributed more than 15% of the country's energy demand [32]. Alleviating the demand for fossil fuels and creating jobs in rural areas, benefiting local communities. The sugar and alcohol sector directly employs more than 733,000 people in Brazil [70], with a significant portion of these jobs linked to the use of waste for energy generation. Beyond energy production, there is a wide range of applications in sectors such as manufacturing organic fertilizers, biodegradable plastics, and sustainable construction materials (Fig. 2) [26, 65]. The use of rice husks, for example, in producing eco-bricks or reinforced cement, is an innovation that can generate jobs in urban and rural areas while reducing the construction industry's carbon footprint [77].

However, while the opportunities are promising, the challenges are also significant. Small and medium-sized producers often generate most of the agro-industrial waste and need help accessing advanced waste processing technologies [28]. Many of these farmers need more financial resources or technical knowledge to adopt waste valorization practices effectively. This scenario could create a gap between large corporations, which can invest in cutting-edge technologies, and small producers,

Fig. 1 Quantity of waste generated annually along the food production chain in 2022. *Source* Authors (2024)

who need to be included in innovation opportunities. To overcome these challenges, public policies must be implemented to facilitate access to sustainable technologies. Governments can play a crucial role by providing subsidies, tax incentives, and accessible credit lines to encourage the adoption of waste valorization practices.

Additionally, the creation of regional processing centers, functioning as technological hubs, could connect small farmers to new opportunities and allow them to benefit from the circular economy. The Agri Frontier Growth Hub, for example, has been assisting farmers in Africa with technical support and financing, making the implementation of sustainable practices more feasible [4]. These initiatives provide resources, knowledge, and even subsidies, promoting the circular economy in a practical and scalable way for small producers.

Another crucial socioeconomic aspect is social inclusion. Large corporations dominate agricultural industries in many world regions, controlling vast land areas

Fig. 2 Possible uses of agro-industrial waste. *Source* Authors (2024)

and most financial resources [21]. Creating a scenario in which small farmers and rural workers have fewer opportunities to participate in the economic benefits generated by waste valorization. However, when well-implemented, agro-industrial waste valorization can have the opposite effect: it can promote social inclusion by democratizing access to new value chains. Agricultural cooperatives, for example, have proven to be an effective solution to overcome entry barriers to waste valorization technologies [2, 75]. By joining forces, small producers can share the investment costs in infrastructure and technologies, making them more accessible. Additionally, these cooperatives can play a central role in negotiating better prices for products derived from waste, strengthening the bargaining power of small farmers in the market [22].

From an economic perspective, the impact of waste valorization can be even more significant when considering the possibility of transforming agro-industrial by-products into ingredients for the pharmaceutical industry. Bioactive compounds, antioxidants, anti-inflammatory, and antibacterial substances, often present in peels, leaves, and other plant residues, can be extracted and used in the production of medicines and nutritional supplements [48, 91]. Creating a new frontier for innovation, where discarded waste becomes a high-value input, generating new economic opportunities for industries and rural communities.

The production of pharmaceuticals from agro-industrial waste is a growing field with significant potential to create markets and job opportunities. Imagine that fruit peels, often discarded, contain compounds with therapeutic properties that could be used to treat chronic diseases. However, a collaborative environment between industrial sectors, research institutions, and governments is essential for these advancements to have a positive socioeconomic impact. Creating policies that incentivize innovation and applied research in biotechnology can accelerate the discovery of new applications for agro-industrial waste while promoting the economic development of rural communities. Strategic partnerships between private companies and public institutions are crucial for translating these discoveries from the lab into viable market solutions.

Globally, the valorization of agro-industrial waste also has the potential to transform the economies of developing countries, particularly in Latin America, Africa, and Asia. In these regions, where agriculture represents a significant portion of GDP, waste can drive economic and social development. Therefore, agro-industrial waste valorization is much more than an environmental solution. It is a transformative socioeconomic opportunity, capable of creating jobs, improving the income of small farmers, promoting social inclusion, and opening new frontiers for innovation. Looking at agro-industrial waste from a new perspective, we see that it is not just a by-product of a production process. They are resources with untapped potential, ready to drive a new era of sustainable development and economic inclusion.

3 Valorization of Agro-industrial Waste: Technical Challenges and Possible Products

Agro-industrial waste originates from organic materials from processes or materials from agriculture, livestock, forestry, and processing and transformation industries [88]. Leaves, stems, fruit parts and seeds are examples of agro-industrial waste from different crops in the agricultural sector, such as sugar cane, corn, soybeans, rice, cotton, and oranges, among others [52, 69, 86]. In the livestock sector, agro-industrial waste is usually manure and leftovers from meat processing, with the use of bones standing out [54, 99]. In the forestry sector, agro-industrial waste includes leaves, branches, bark, log ends, and materials from wood processing [95].

In addition to the large quantities of material generated from different raw material sources and their heterogeneous characteristics, agro-industrial waste can cause serious social, environmental, and economic problems if it is not managed correctly. An excellent alternative for using agro-industrial waste is to valorize it through biological, physical, chemical, and thermal processes to generate energy and bioproducts [71, 102]. A more up-to-date alternative for the use and recovery of agro-industrial waste is its use in biorefinery facilities, where all the material used is transformed into a high-value-added bioproduct or energy.

3.1 Bioproducts

Bioproducts generated from the conversion of agro-industrial waste are diverse, such as pigments and coagulants [11, 38, 72]. These materials can be used in the food, pharmaceutical, cosmetics, varnish, paint, and synthetic materials sectors [16, 41, 64].

3.1.1 Pigments

Natural pigments are predominantly derived from fruits, vegetables, microorganisms, food waste, and materials from food processing industries [43]. The main pigments extracted from agro-industrial waste are anthocyanins, tannins, carotenoids, and chlorophylls [56].

Anthocyanins are water-soluble natural dyes belonging to the flavonoid class, capable of providing red, purple, or blue colors, depending on their chemical composition [35]. Anthocyanins are commonly found in grapes, jaboticabas, beets, apples, and blueberries [30, 44]. Anthocyanin compounds are formed from a basic flavylium cation structure ($C_{15}H_{11}O^+$) (Table 1) with seven radicals that can be replaced by hydrogen (H), hydroxyls (OH) or methoxyls (OCH_3) [57]. Anthocyanins have several properties that can benefit human health, as they are chemical components with antioxidant and anti-inflammatory characteristics and can contribute to lowering blood pressure [9, 56].

Tannins are natural compounds that can act as natural colorants, providing a brownish color or as flavor enhancers, providing a sensation of astringency in food [58]. Tannins can be divided into hydrolyzable, found in fruits and plants, and water-soluble and condensed, found more frequently in tree bark, being more resistant to decomposition [8, 53]. In the leather production industry, tannins are used in the tanning process, which turns animal skin into a high-value-added product with hydrothermal properties [55]. In the drinks industry, tannins are present in most red wines, acting as natural preservatives, giving the drink a longer shelf life [100].

Carotenoids are lipophilic natural colorants with yellowish, orange, and reddish hues. They are found in carrots, tomatoes, corn, spinach, and cabbage [56]. These compounds are formed from isoprene units (C_5H_8) linked in long chains of conjugated double bonds [59]. There are more than 600 different types of carotenoids, however, the best known are β-carotene ($C_{40}H_{56}$), lycopene ($C_{40}H_{56}$) and lutein ($C_{40}H_{56}O_2$) (Fig. 3), as they are used in health because they are precursors of vitamins

Table 1 The most common anthocyanins in nature

Base structure	Compound name	R1	R2	R3	R4	R5	R6	R7
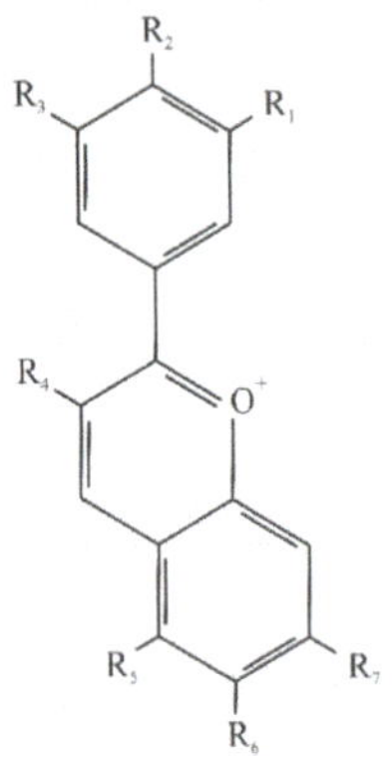	Cyanidin	OH	OH	H	OH	OH	H	OH
	Delphinidin	OH	OH	OH	OH	OH	H	OH
	Pelargonidin	H	OH	H	OH	OH	H	OH
	Malvidin	OCH_3	OH	OCH_3	OH	OH	H	OH
	Peonidin	OCH_3	OH	H	OH	OH	H	OH
	Petunidin	OCH_3	OH	OH	OH	OH	H	OH

Source Zang et al. [104]

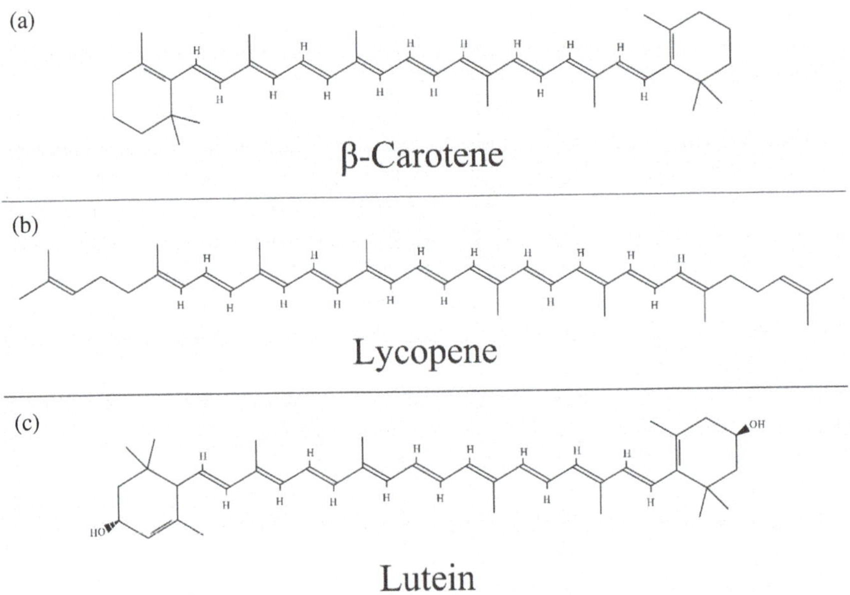

Fig. 3 2-D schematic representation of the main carotenoids: **a** β-carotene, **b** Lycopene, and **c** Lutein. *Source* NCBI [59]

and have anticarcinogenic characteristics, in addition to their use as food additives [1, 17, 97].

Chlorophyll is a natural green pigment found in plants, algae, and some microorganisms. It is essential in photosynthesis, capturing sunlight and converting carbon dioxide and water into sugars and oxygen [60]. There are six types of chlorophyll; however, only two of them are used in the food, cosmetics, and pharmaceutical industries: chlorophyll A ($C_{55}H_{72}MgN_4O_5$) and chlorophyll B ($C_{55}H_{70}MgN_4O_6$) [56]. The main use of chlorophyll in these industries is due to its antioxidant characteristics, inhibiting the oxidation of cells by controlling the production of free radicals [67].

3.1.2 Coagulants

Wastewater treatment is an economically demanding process and is becoming a significant challenge for city managers as populations and consumption of raw materials increase. One of the stages of wastewater treatment is the coagulation process, but its efficiency varies according to the material used as a coagulant. Usually, chemical coagulants are used, which are expensive and not easy to recover for reuse [40]. In this sense, there is a need to use bio-coagulants that are environmentally friendly, economically attractive, and technologically efficient. Several studies have pointed to some agro-industrial waste as examples of raw materials for extracting bio-coagulants, such as guarana [78], sugar cane bagasse [5], orange seeds [47], peanut shells [90], among others. In the context presented above, agro-industrial waste is an excellent source of bio-coagulants, helping to reduce the costs of wastewater treatment and increasing the sustainability of the process.

3.2 Energy

In addition to transforming agro-industrial waste into value-added bioproducts, this material can be converted into energy, be it thermal energy, electrical energy, or products used in energy generation, such as ethanol and biodiesel.

3.2.1 Biological Processes

Biological processes for converting agro-industrial waste into energy usually take place on an industrial scale. Fermentation transforms organic matter into bioethanol or biogas.

The production of ethanol from biomass is already widely known, mainly using sugar cane and corn, in which the raw material is transformed into sugars and subsequently broken down into ethanol (C_2H_6O) and carbon dioxide (CO_2) by yeasts in an alcoholic fermentation process [23]. When it comes to producing ethanol from agro-industrial waste, the scenario changes completely, as the raw material is heterogeneous, which makes it challenging to use on a large scale.

Producing biogas from biomass involves breaking down organic matter into gases rich in methane gas (CH_4) and carbon dioxide (CO_2) in an atmosphere devoid of oxygen, a process called anaerobic digestion. Unlike alcoholic fermentation, anaerobic digestion does not have as many restrictions on the raw material, which makes agro-industrial waste an excellent solution for generating energy from biogas [84].

When evaluating biological processes for converting biomass into energy, biogas production is the most suitable solution, as there are fewer technological barriers to using agro-industrial waste than generating ethanol.

3.2.2 Physical Processes

The physical processes for transforming agro-industrial waste into energy can be divided into two groups: densification (briquette) or extrusion (pellets), which generate two energy products, briquettes and pellets, respectively.

Briquettes are produced from ground biomass inserted into a hydraulic press, with or without a heating stage, depending on the biomass used in the process; the result is a material of reduced volume and high density but with chemical properties similar to the raw material [3]. Briquetting is a complex process to apply to agro-industrial waste, as the material needs to be homogeneous, with a low particle size (< 0.5 mm), low humidity (8–12%), and there may be a need to add binders to the process, which can reduce the cost-effectiveness of the product.

Pellets are produced from ground biomass that passes through an extrusion die, either an extruder disk or a continuous screw system, heated to temperatures between 90 and 120 °C so that the material's lignin enters a plastic regime, acting as a binder [68]. The general specifications for pellet production are regulated by the ISO 17225 series of international standards, in which the density, durability, and lower heating value of the pellets must be greater than or equal to 600 kg m^{-3}, 97.5%, and 16.5 MJ kg^{-1}, respectively, and the moisture and ash content must be less than or equal to 10% and 2%, respectively [42]. Pelletizing presents the same application problems for agro-industrial waste as briquetting, requiring a more homogeneous material before starting the process.

Physical processes for converting agro-industrial waste into energy generally require homogeneous raw materials for greater efficiency. However, investments in research and innovation are being made to use these processes in conjunction with biological, thermochemical, and biorefinery processes to obtain better energy yields.

3.2.3 Thermochemical Processes

Thermochemical processes for converting agro-industrial waste into energy can be divided into two large groups: exothermic and endothermic. The best-known exothermic process is combustion, which consists of the thermochemical reaction between an organic material (fuel) and oxygen (oxidizer), releasing energy, water, and gases, depending on the composition of the fuel and whether the combustion process was complete or incomplete.

Endothermic processes receive energy to transform organic raw materials into products with a greater amount of energy released compared to the source material [96]. Depending on the process's characteristics or the desired product, endothermic processes can be divided into torrefaction, pyrolysis, and gasification.

Torrefaction is a thermochemical process for modifying biomass that takes place at temperatures between 200 and 300 °C in an inert atmosphere (N_2 or H_2), generating three materials: gases (CO, CO_2, CH_4, H_2, and C_xH_y), liquids (H_2O, acetic acid, alcohols, and aldehydes) and solids (torrefied material). Torrefied material differs from raw biomass in some characteristics: low humidity, higher calorific value, lower oxygen/carbon (O/C) and hydrogen/carbon (H/C) ratios, it becomes a hydrophobic material and is easier to grind [62]. Depending on how the torrefaction process is carried out, especially about the final temperature, it can be classified into three types of process: mild (200–235 °C), moderate (235–275 °C) or severe (275–300 °C), increasing the energetic properties of the material as the final temperature increases [24]. In using agro-industrial waste for energy generation, torrefaction can be used as a pre-treatment of the raw material to homogenize and improve the physical–chemical characteristics of this waste compared to its natural state.

Pyrolysis is a thermochemical process for modifying biomass that takes place between temperatures of 300 and 700 °C in an atmosphere with little oxygen or an inert atmosphere, resulting in three products: gases (CO, CO_2, CH_4, H_2, and C_xH_y), liquids (pyroligneous liquid) and solids (charcoal). Pyrolysis can be classified according to the process's final temperature, heating rate, residence time, and desired product. Slow pyrolysis takes place over hours or even days, with a very low heating rate (< 1 °C min^{-1}) and a final temperature of over 400 °C, intending to produce charcoal [14]. Fast and flash pyrolysis occur with residence times of less than five seconds, with high heating rates (hundreds of °C min^{-1} and thousands of °C min^{-1}, respectively) and final temperatures between 500 and 700 °C, to obtain pyroligneous liquid and chemical compounds [50]. For fast or flash pyrolysis to obtain the highest yield of pyroligneous liquid, the process must be carried out in a fluidized bed with a gas cooling system, favoring the liquid formation route and avoiding the degradation of organic vapors [63]. For agro-industrial waste, pyrolysis can be used at the household level to generate charcoal or on an industrial scale to generate energy for systems or to produce bioproducts from the fractionation of pyroligneous liquid.

Gasification is a thermochemical process that occurs at temperatures above 700 °C in mobile reactors (rotary or continuous flow), in an atmosphere of oxygen, flow of vapors (water), or atmospheric air, generating synthesis gas at the end of the process. Gasification takes place in three stages: (I) Drying, in which the biomass loses a large amount of water; (II) Pyrolysis, producing charcoal and releasing large quantities of gases; (III) Gasification, when the charcoal breaks down rapidly, creating synthesis gas (a mixture of gases rich in H_2 and CO), used in the bioproducts or renewable fuels sectors [15]. Agro-industrial waste can go through the gasification process at an industrial level when using the concept of biorefineries, where part of the waste is used for heating and part is directed to the reactor to generate synthesis gas.

3.3 *Biorefineries*

There are several concepts for biorefineries, and most highlight critical aspects, such as a highly efficient biomass processing system, converting all organic matter into energy or some high-value-added material. According to the International Energy Agency, biorefineries are "the sustainable processing of biomass into a spectrum of commercial products and energy" [25]. Biorefineries can be used in various processes for different types of materials, such as generating biogas by using food waste [46], using waste from industrial processes [49] or in large production plants [10]. Because it is a "closed system" concept in which all the raw material that enters the process is transformed into a product or energy at the end with little or no loss, biorefineries are an excellent alternative for using agro-industrial waste (Fig. 4).

The scale of a biorefinery for using this waste depends on several factors, including the homogeneity of the raw material, the type of agro-industrial activity carried out, and the yield in product or energy at the end of the process. For example, when dealing with residues from agro-industrial activities that involve more than one crop, the most viable option is to use micro-biorefineries for each residue. At the same time, a large-scale biorefinery is the most suitable option for pulp production plants.

4 Challenges, Opportunities, and Prospects Regarding the Recovery of Agro-industrial Waste

4.1 *Prospects and Opportunities*

Based on the evidence and discussions presented throughout this book, the valorization of agro-industrial waste represents a viable and promising alternative for developing new bioproducts. As technologies advance and the needs of the pharmaceutical industry become more demanding, the exploitation of waste as a source of bioactive compounds is emerging as a central strategy for sustainability in the manufacture of herbal products. Given this scenario, it is pertinent to explore the main opportunities and challenges for the future of this field, especially concerning the production of herbal medicines and the integration between the agro-industrial and pharmaceutical sectors.

The possibilities for using agro-industrial waste encompass different possibilities and challenges. The inclusion of biorefineries in the agro-industrial sector is an alternative that allows the sustainable and efficient extraction of different chemical compounds. In a biorefinery, waste from the agricultural sector can be transformed into various types of products, from biofuels to bioactive compounds, thus aiming for a complete cycle of material utilization and reducing the consumption of non-renewable sources.

For this utilization system to be fully realized, it is essential to form strategic partnerships and invest in joint research. Collaboration between governments, research

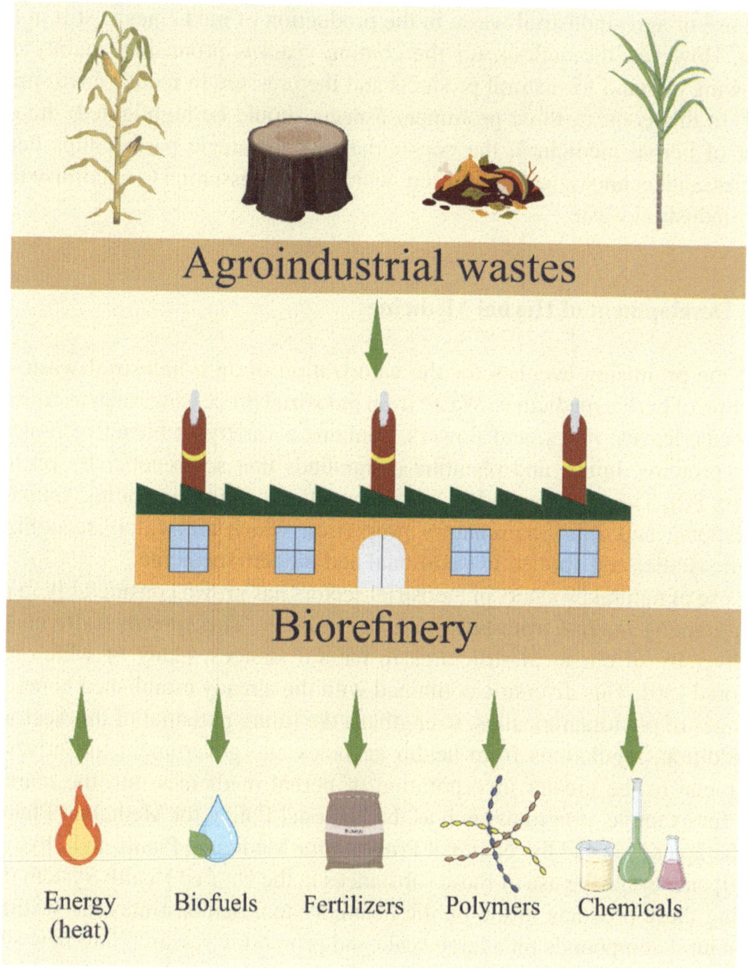

Fig. 4 Use of agro-industrial wastes in biorefineries for energy, biofuels, and bioproducts generation. *Source* Authors (2024)

groups, educational institutions, and companies in general is essential to consolidate the use of this agro-industrial waste in various industries, such as the energy, food, and pharmaceutical sectors, on a global scale.

In addition to these partnerships, technological innovation is one of the main reasons for valorizing agro-industrial waste. Advances in techniques for extracting, purifying, and formulating new pharmaceutical agents from natural sources have grown significantly, making the possibility of using agro-industrial waste as a raw material a reality. In recent years, interest in natural and sustainable products has grown, and technologies and research have been vital to ensuring that these by-products meet the parameters of sustainable products.

The use of agro-industrial waste in the production of medicines is still in development. However, the outlook for the coming years is promising, mainly due to the growing demand for natural products and the pressure to reduce environmental impact. In this context, three prominent aspects should be highlighted: the development of herbal medicines, the consolidation of strategic partnerships between industries, and technological innovation, which will be essential to maximize the use of agro-industrial waste.

4.1.1 Development of Herbal Medicines

One of the promising avenues for the valorization of agro-industrial waste is the production of herbal medicines. Waste from industrial processing, such as fruit, pulp, peel, seeds, leaves, twigs, and flowers, contains a variety of bioactive molecules, such as proteins, lipids, and phenolic compounds that act beneficially on human health [6, 80]. These compounds have therapeutic potential, including antioxidant, antimicrobial, and anti-inflammatory properties, which are widely recognized in academic studies and applied in traditional and modern medicine.

The use of natural products in industrial sectors has grown considerably in recent years, attracting interest from academia and industry. This growth is driven by the great diversity of chemical structures in natural sources, many of which remain unexplored [36]. This diversity, combined with the already established benefits and advantages of phytotherapeutics, strengthens the future potential of this sector.

In addition, regulations from health agencies and government incentives have been crucial to the greater incorporation of herbal medicines into the market. In Brazil, for example, programs such as the National Policy for Medicinal Plants and Phytotherapics [19] and the National Program for Medicinal Plants and Phytotherapics [20] encourage the use of these substances in the Unified Health System (SUS). Including these products in the public health system demonstrates the viability of using natural compounds on a large scale and promoting research into new sources of medicines, especially those derived from industrial waste that contains bioactive compounds.

4.1.2 Strategic Partnerships and Technological Innovation

Strengthening partnerships between the agro-industrial and pharmaceutical sectors is an essential strategy for expanding the use of agro-industrial waste in producing medicines. The sustainability of this process depends directly on global access to information about plants and their compounds, as well as collaborative international efforts involving governments, academic institutions, non-governmental organizations, and the private sector [66]. These cross-sector collaborations allow for forming more integrated and efficient production chains in which the waste generated by one industry can be reused as inputs by another, thus promoting an actual circular economy.

Long-term sustainability, meanwhile, requires the implementation of responsible cultivation practices and the conscious use of bioeconomy resources, permanently preserving biodiversity. It is necessary to establish transparent commercial partnerships where all parties involved—from producers to end consumers—can obtain clear and balanced benefits [45]. In this context, the agroindustry is vital as a supplier of alternative and sustainable raw materials. At the same time, the pharmaceutical industry, in turn, benefits from exploring new bioactive molecules of high therapeutic value found in these residues.

In recent years, initiatives to create biorefineries have intensified, where cutting-edge technologies such as supercritical CO_2 extraction are applied to obtain bioactive compounds from agro-industrial waste in a highly efficient manner [51]. These industrial plants, which are integrated into value chains linking agroindustry and the pharmaceutical industry, demonstrate the viability of innovative solutions for reusing waste on a large scale. In addition to promoting sustainability, these initiatives also show the potential to generate significant economic benefits.

On the global stage, the formation of strategic partnerships and investment in collaborative research are essential factors for progress in this sector. Developing countries with rich biodiversity and extensive agro-industrial chains have untapped potential to become leaders in producing medicines from waste [13]. However, joint actions between governments, research institutions, and private companies are essential for this potential to be fully realized. These collaborations will be decisive in boosting the recovery of agro-industrial waste on a global scale, consolidating the role of waste as a sustainable source of raw materials for the pharmaceutical industry.

It is essential to combine technological progress with the formation of strategic partnerships and collaborative research to valorize agro-industrial waste and develop new pharmaceutical agents more fluidly. In a global scenario, investment in technological innovation such as extraction, purification, and drug formulation techniques increase the viability of using this waste as raw material in the pharmaceutical industry. In recent years, natural products have played a vital role in the development of new drugs, and this contribution continues to be relevant.

Although the isolation of pure natural products from complex mixtures remains a challenge—especially considering that we do not yet have single-step isolation procedures—the application of more selective methods, ranging from extraction to fractionation and purification, promises to speed up the process from the collection of biological material to the isolation of the final purified compound [103]. These advanced techniques enable the extraction of compounds in high concentrations and with high levels of purity, essential characteristics to ensure the safety and efficacy of medicines developed from agro-industrial waste.

4.1.3 Sustainability and Global Impact

The use of agro-industrial waste in the manufacture of medicines is intrinsically aligned with the principles of sustainability and the circular economy. The concept of the circular economy is based on the idea that the waste generated by one industry can serve as raw materials for another, thus promoting a more efficient use of available resources. This model challenges the classic linear paradigm of the economy, which boils down to making, using, and disposing by incorporating the principles of reduction, reuse, recycling, recovery, and restoration, also known as the 5Rs of sustainable development [83]. By adopting this approach, industries minimize waste and create opportunities for innovation and value creation.

The implementation of the circular bioeconomy concept, therefore, emerges as a promising strategy for waste management, contributing to the solution of various global problems facing society today [79]. In this context, the recovery of agro-industrial waste is fundamental, as it is aligned with the Sustainable Development Goals (SDGs) established by the United Nations. These goals highlight the importance of sustainable practices and include specific targets related to bioeconomy, recognizing the vital role of bio-based resources in building a sustainable future [98]. SDG 3, which aims to promote health and well-being, and SDG 12, which seeks to ensure responsible consumption and production patterns, emphasize the relevance of this interconnection [92].

Furthermore, promoting a greener pharmaceutical industry, which adopts sustainable practices at all stages—from the selection of raw materials to the final production of medicines—not only contributes to the preservation of the environment but also improves people's quality of life. By integrating circular economy and bioeconomy principles into its operations, the pharmaceutical industry can be crucial in mitigating environmental impacts while meeting the growing demand for effective and safe products. Thus, the recovery of agro-industrial waste is an opportunity for innovation and an ethical and strategic imperative in pursuit of more sustainable development.

The prospects for recovering agro-industrial waste in the manufacture of medicines are undoubtedly promising. As technologies advance, the potential for harnessing waste as a source of bioactive compounds grows exponentially. Partnerships between the agro-industrial and pharmaceutical sectors, combined with new extraction techniques and technological innovation, will drive the development of more efficient, safe, and sustainable medicines.

For this potential to be fully realized, greater integration between the research and industry sectors and the development of public policies that encourage the reuse of waste on a global scale must occur. By making more efficient use of available resources, the pharmaceutical industry can transform itself, becoming not only more sustainable but also capable of providing innovative solutions to contemporary society's health challenges.

5 Conclusions

When viewed as a whole, without dividing it into sectors, the agroindustry is an excellent source of raw materials for generating high-value-added bioproducts or renewable energy. However, several socioeconomic and technological barriers prevent the use of agro-industrial waste, making some processes unviable for immediate implementation.

Since organic matter is generated in large quantities and could cause serious socio-environmental problems due to lack of proper disposal, agro-industrial waste is an alternative and a solution for producing various bioproducts (pigments, coagulants, enzymes) or energy through various processes. In this sense, agro-industrial waste is in line with Sustainable Development Goals (SDGs) 3, 7, 12, and 13, becoming a practice for the future of the world's communities. The prospects for using agro-industrial waste more assertively are the development of research into using this material on any scale, whether domestic or industrial, contributing to social, economic, and environmental development.

References

1. Abdel-Aal ESM, Akhtar H, Zaheer K, Ali R (2013) Dietary sources of lutein and zeaxanthin carotenoids and their role in eye health. Nutrients 5:1169–1185
2. Ada E, Sagnak M, Uzel RA, Balcıoğlu İ (2022) Analysis of barriers to circularity for agricultural cooperatives in the digitalization era. Int J Prod Perform Manage 71(3):932–951. https://doi.org/10.1108/IJPPM-12-2020-0689/FULL/PDF
3. Adam R, Yiyang D, Kruggel-Emden H, Zeng T, Lenz V (2024) Influence of pressure and retention time on briquette volume and raw density during biomass densification with an industrial stamp briquetting machine. Renew Energy 229. https://doi.org/10.1016/j.renene.2024.120773
4. Agri Frontier (2024) From waste to wealth: transforming agri by-products into value. https://www.agrifrontier.com/case-study/waste-to-wealth/. Accessed 29 Oct 2024
5. Ahmed HM, El-Khateeb MA, Mohamed NY, Sobhy NA, Fawzy ME (2024) Evaluation of different natural waste materials as bio-coagulants for domestic wastewater treatment. Desalin Water Treat 317. https://doi.org/10.1016/j.dwt.2024.100034
6. Aili Q, Cui D, Li Y, Zhige W, Yongping W, Minfen Y, Dongbin L, Xiao R, Qiang W (2024) Composing functional food from agro-forest wastes: selectively extracting bioactive compounds using supercritical fluid extraction. Food Chem 455:139848. https://doi.org/10.1016/J.FOODCHEM.2024.139848
7. Andrade CR, Cupertino GFM, da Silva ÁM, Brito JO, Morais WWC, Balboni BM, Saloni D, Dias Júnior AF (2024) The potential of wood-based urban waste to generate bioenergy and increase the energetic sustainability. Clean Technol Environ Policy 26(9):2885–2898. https://doi.org/10.1007/S10098-024-02775-5/FIGURES/7
8. Arapitsas P (2012) Hydrolyzable tannin analysis in food. Food Chem 135:1708–1717
9. Assous MTM, Abdel-Hady MM, Medany GM (2014) Evaluation of red pigment extracted from purple carrots and its utilization as antioxidant and natural food colorants. Ann Agric Sci 59(1):1–7. https://doi.org/10.1016/j.aoas.2014.06.001
10. Aui A, Wang Y (2023) Cellulosic ethanol production: assessment of the impacts of learning and plant capacity. Technol Forecast Soc Change 197. https://doi.org/10.1016/j.techfore.2023.122923
11. Avelino Gonçalves F, Dos Santos ES, De Macedo GR (2015) Use of cultivars of low cost, agro-industrial and urban waste in the production of cellulosic ethanol in Brazil: a proposal to utilization of microdistillery. Renew Sustain Energy Rev 50:1287–1303
12. Babu S, Singh Rathore S, Singh R, Kumar S, Singh VK, Yadav SK, Yadav V, Raj R, Yadav D, Shekhawat K, Ali Wani O (2022) Exploring agricultural waste biomass for energy, food and

feed production and pollution mitigation: a review. Bioresour Technol 360:127566. https://doi.org/10.1016/J.BIORTECH.2022.127566

13. Banwo K, Olojede AO, Adesulu-Dahunsi AT, Verma DK, Thakur M, Tripathy S, Singh S, Patel AR, Gupta AK, Aguilar CN, Utama GL (2021) Functional importance of bioactive compounds of foods with potential health benefits: a review on recent trends. Food Biosci 43:101320. https://doi.org/10.1016/J.FBIO.2021.101320
14. Basu P (2018) Pyrolysis. In: Biomass gasification, pyrolysis and torrefaction: practical design and theory. Elsevier, pp 155–187
15. Basu P, Kaushal P (2024) Gasification theory. In: Biomass gasification, pyrolysis, and torrefaction. Elsevier, pp 207–258
16. Benito-González I, del Mar Ortiz-Gimeno M, López-Rubio A, Martínez-Abad A, Garrido-Fernández A, Martínez-Sanz M (2022) Sustainable starch biocomposite films fully-based on white rice (Oryza sativa) agro-industrial by-products. Food Bioprod Process 136:47–58. https://doi.org/10.1016/j.fbp.2022.09.008
17. Bora P, Das P, Bhattacharyya R, Barooah MS, Saikia Barooah M (2019) Biocolour: the natural way of colouring food. J Pharmacognosy Phytochem 8(3):3663–3668
18. Brasil (2024) DECRETO No 12.082, DE 27 DE JUNHO DE 2024 - Institui a Estratégia Nacional de Economia Circular. https://www.planalto.gov.br/ccivil_03/_ato2023-2026/2024/decreto/D12082.htm. Accessed 29 Oct 2024
19. Brasil. Ministério da Saúde (MS) (2006). Política Nacional de Plantas Medicinais e Fitoterápicos. MS, Brasília (Série B. Textos Básicos de Saúde)
20. Brasil. Ministério da Saúde (MS) (2009). Programa Nacional de Plantas Medicinais e Fitoterápicos. MS, Brasília (Série C. Projetos, Programas e Relatórios)
21. Burja V, Tamas-Szora A, Dobra IB (2020) Land concentration, land grabbing and sustainable development of agriculture in Romania. Sustainability 12(5):2137. https://doi.org/10.3390/SU12052137
22. Candemir A, Duvaleix S, Latruffe L (2021) Agricultural cooperatives and farm sustainability—a literature review. J Econ Surv 35(4):1118–1144. https://doi.org/10.1111/JOES.12417
23. Cardona CA, Quintero JA, Paz IC (2010) Production of bioethanol from sugarcane bagasse: status and perspectives. Bioresour Technol 101(13):4754–4766. https://doi.org/10.1016/j.biortech.2009.10.097
24. Chen WH, Peng J, Bi XT (2015) A state-of-the-art review of biomass torrefaction, densification and applications. Renew Sustain Energy Rev 44:847–866
25. Cherubini F, Mandl M, Philips C, Wellisch M, Jørgensen H, Skiadas IV, Boniface L, Dohy M, Pouet J-C, Wilke T, Walsh P, Van Ree R, Jong E (2010) General rights IEA bioenergy Task 42 on biorefineries: co-production of fuels, chemicals, power and materials from biomass IEA bioenergy Task 42-Countries Report production of fuels, chemicals, power and materials from biomass: IEA bioenergy Task 42-Countries Report. IEA Bioenergy. APA
26. Chhetri RK, Aryal N, Kharel S, Chandra Poudel R, Pant D (2020) Agro-based industrial wastes as potent sources of alternative energy and organic fertilizers. Curr Dev Biotechnol Bioeng Sustain Bioresour Emerg Bioeconomy 121–136. https://doi.org/10.1016/B978-0-444-64309-4.00005-2
27. Chilakamarry CR, Mimi Sakinah AM, Zularisam AW, Sirohi R, Khilji IA, Ahmad N, Pandey A (2022) Advances in solid-state fermentation for bioconversion of agricultural wastes to value-added products: opportunities and challenges. Bioresour Technol 343:126065. https://doi.org/10.1016/J.BIORTECH.2021.126065
28. Derhab N, Elkhwesky Z (2023) A systematic and critical review of waste management in micro, small and medium-sized enterprises: future directions for theory and practice. Environ Sci Pollut Res 30(6):13920–13944. https://doi.org/10.1007/S11356-022-24742-7/FIGURES/1
29. Dey P, Gupta G, Nayak J, Dilip KJ (2022) Improved enzymatic hydrolysis of lignocellulosic waste biomass: most essential stage to develop cost-effective second-generation biofuel production. In: Biofuels and bioenergy. Elsevier, pp 133–165

30. Eker ME, Aaby K, Budic-Leto I, Brncic SR, El SN, Karakaya S, Simsek S, Manach C, Wiczkowski W, De Pascual-Teresa S (2020) A review of factors affecting anthocyanin bioavailability: possible implications for the inter-individual variability. Foods 9(1). https://doi.org/10.3390/foods9010002
31. El Sheikha AF, Ray RC (2023) Bioprocessing of horticultural wastes by solid-state fermentation into value-added/innovative bioproducts: a review. Food Rev Intl 39(6):3009–3065. https://doi.org/10.1080/87559129.2021.2004161
32. EPE (2023) [Empresa de Pesquisa Energética] Brazilian Energy Balance 2023 year 2022. Rio de Janeiro
33. European Commission (2023) Circular economy action plan—European Commission. https://environment.ec.europa.eu/strategy/circular-economy-action-plan_en#documents. Accessed 29 Oct 2024
34. Fytili D, Zabaniotou A (2022) Organizational, societal, knowledge and skills capacity for a low carbon energy transition in a circular waste bioeconomy (CWBE): observational evidence of the Thessaly region in Greece. Sci Total Environ 813:151870. https://doi.org/10.1016/J.SCITOTENV.2021.151870
35. Ghosh S, Sarkar T, Das A, Chakraborty R (2022) Natural colorants from plant pigments and their encapsulation: an emerging window for the food industry. LWT 153
36. Gonçalves S, Romano A (2021) Green approaches for the extraction of bioactives from natural sources for pharmaceutical applications. Green Sustain Process Chem Environ Eng Sci Solvents Pharm Ind 249–267. https://doi.org/10.1016/B978-0-12-821885-3.00013-X
37. Gonzalez-Arcos C, Joubert AM, Scaraboto D, Guesalaga R, Sandberg J (2021) "How do I carry all this now?" Understanding consumer resistance to sustainability interventions. J Mark 85(3):44–61. https://doi.org/10.1177/0022242921992052/ASSET/IMAGES/LARGE/10.1177_0022242921992052-FIG7.JPEG
38. Greses S, Tomás-Pejó E, Gónzalez-Fernández C (2020) Agro-industrial waste as a resource for volatile fatty acids production via anaerobic fermentation. Bioresour Technol 297. https://doi.org/10.1016/j.biortech.2019.122486
39. Gupta J, Kumari M, Mishra A, Swati AM, Thakur IS (2022) Agro-forestry waste management—a review. Chemosphere 287:132321. https://doi.org/10.1016/J.CHEMOSPHERE.2021.132321
40. Hellal MS, Rashad AM, Kadimpati KK, Attia SK, Fawzy ME (2023) Adsorption characteristics of nickel (II) from aqueous solutions by Zeolite Scony Mobile-5 (ZSM-5) incorporated in sodium alginate beads. Sci Rep 13(1). https://doi.org/10.1038/s41598-023-45901-x
41. Hero JS, Romero CM, Pisa JH, Perotti NI, Olivaro C, Martinez MA (2018) Designing cross-linked xylanase aggregates for bioconversion of agro-industrial waste biomass towards potential production of nutraceuticals. Int J Biol Macromol 111:229–236. https://doi.org/10.1016/j.ijbiomac.2017.12.166
42. International Organization for Standardization (ISO) (2021) Solid biofuels—fuel specifications and classes. Part 2: graded wood pellets, 2nd edn. ISO, Genebra
43. Kamalesh R, Saravanan A, Yaashikaa PR, Vijayasri K (2025) Innovative approaches to harnessing natural pigments from food waste and by-products for eco-friendly food coloring. Food Chem 463:141519. https://doi.org/10.1016/j.foodchem.2024.141519
44. Khan MK, Zill-E-Huma DO (2014) A comprehensive review on flavanones, the major citrus polyphenols. J Food Compos Anal 33:85–104
45. Kurnaz ML, Aksan Kurnaz I (2021) Commercialization of medicinal bioeconomy resources and sustainability. Sustain Chem Pharm 22:100484. https://doi.org/10.1016/J.SCP.2021.100484
46. Kurniawan KIA, Susanti H, Rani DS, Harahap BM, Firmansyah EA, Ishizaki R, Demura M, Ahamed T, Noguchi R (2024) Techno-economic analysis of biocrude, biogas, and fertilizer production from microalgae Coelastrella striolata cultivated in agro-industrial wastewater. J Clean Prod 477. https://doi.org/10.1016/j.jclepro.2024.143857
47. Licona–Aguilar AI, Torres–Huerta AM, Domínguez–Crespo MA, Negrete–Rodríguez MLX, Conde–Barajas E, Brachetti–Sibaja SB, Rodríguez–Salazar AE (2024) Valorization of agro-industrial orange peel waste during the optimization of activated carbon–multiwalled carbon

nanotubes–zinc oxide composites used in the removal of methylene blue in wastewater. Chem Eng J 492. https://doi.org/10.1016/j.cej.2024.152102
48. Lima RS, de Carvalho APA, Conte-Junior CA (2023) Health from Brazilian Amazon food wastes: bioactive compounds, antioxidants, antimicrobials, and potentials against cancer and oral diseases. Crit Rev Food Sci Nutr 63(33):12453–12475. https://doi.org/10.1080/10408398.2022.2101983
49. Liu Q, Ji X, Zhu K, Wang W, Bao Z, Zhang L (2024) Biorefinery-oriented pyrolysis and anaerobic fermentation synergies: leveraging biomass tar for enhanced chemical production from food waste. J Clean Prod 477. https://doi.org/10.1016/j.jclepro.2024.143830
50. Mahinpey N, Abdalla A, Farooqui A (2023) Biomass gasification for hydrogen production: a pathway to cleaner energy transition. In: Biomass to bioenergy: modern technological strategies for biorefineries. Elsevier, pp 205–235
51. Makepa DC, Chihobo CH (2024) Barriers to commercial deployment of biorefineries: a multi-faceted review of obstacles across the innovation chain. Heliyon 10(12):e32649. https://doi.org/10.1016/J.HELIYON.2024.E32649
52. Martinez-Hernandez E, Amezcua-Allieri MA, Sadhukhan J, Anell JA (2018) Sugarcane bagasse valorization strategies for bioethanol and energy production. In: Sugarcane—technology and research. InTech
53. Mena P, Calani L, Bruni R, Del Rio D (2015) Bioactivation of high-molecular-weight polyphenols by the gut microbiome. In: Diet-microbe interactions in the gut: effects on human health and disease. Elsevier Inc., pp 73–101
54. Mendes KF, Hall KE, Takeshita V, Rossi ML, Tornisielo VL (2018) Animal bonechar increases sorption and decreases leaching potential of aminocyclopyrachlor and mesotrione in a tropical soil. Geoderma 316:11–18. https://doi.org/10.1016/j.geoderma.2017.12.017
55. Mustafa MA, Noyon MAR, Uddin ME, Islam R (2024) Sustainable leather tanning with Pontederia crassipes tannin: a promising eco-friendly alternative. Clean Eng Technol 18. https://doi.org/10.1016/j.clet.2023.100717
56. Nabi BG, Mukhtar K, Ahmed W, Manzoor MF, Ranjha MMAN, Kieliszek M, Bhat ZF, Aadil RM (2023) Natural pigments: anthocyanins, carotenoids, chlorophylls, and betalains as colorants in food products. Food Biosci 52
57. National Center for Biotechnology Information (NCBI) (2024a) PubChem compound summary for CID 145858, Flavylium. https://pubchem.ncbi.nlm.nih.gov/compound/Flavylium. Accessed 30 Oct 2024
58. National Center for Biotechnology Information (NCBI) (2024b) PubChem compound summary for CID 16129778, Gallotannin. https://pubchem.ncbi.nlm.nih.gov/compound/Gallotannin. Accessed 30 Oct 2024
59. National Center for Biotechnology Information (NCBI) (2024c) PubChem compound summary for CID 6557, Isoprene. https://pubchem.ncbi.nlm.nih.gov/compound/Isoprene. Accessed 30 Oct 2024
60. National Center for Biotechnology Information (NCBI) (2024d) PubChem compound summary for CID 12085802, Chlorophyll A. https://pubchem.ncbi.nlm.nih.gov/compound/12085802. Accessed 30 Oct 2024
61. Nair LG, Agrawal K, Verma P (2022) An overview of sustainable approaches for bioenergy production from agro-industrial wastes. Energy Nexus 6:100086. https://doi.org/10.1016/J.NEXUS.2022.100086
62. Nunes LJR, De Oliveira Matias JC, Da Silva Catalão JP (2018) Biomass torrefaction process. In: Torrefaction of biomass for energy applications. Elsevier, pp 89–124
63. Nzihou A, Stanmore B, Lyczko N, Minh DP (2019) The catalytic effect of inherent and adsorbed metals on the fast/flash pyrolysis of biomass: a review. Energy 170:326–337
64. Ortiz-Sanchez M, Omarini AB, González-Aguirre JA, Baglioni M, Zygadlo JA, Breccia J, D'Souza R, Lemesoff L, Bodeain M, Cardona-Alzate CA, Pejchinovski I, Fernandez-Lahore MH (2023) Valorization routes of citrus waste in the orange value chain through the biorefinery concept: the Argentina case study. Chem Eng Process Process Intensification 189. https://doi.org/10.1016/j.cep.2023.109407

65. Peng X, Jiang Y, Chen Z, Osman AI, Farghali M, Rooney DW, Yap PS (2023) Recycling municipal, agricultural and industrial waste into energy, fertilizers, food and construction materials, and economic feasibility: a review. Environ Chem Lett 21(2):765–801. https://doi.org/10.1007/S10311-022-01551-5
66. Pešić M (2015) Development of natural product drugs in a sustainable manner. Brief for United Nations Global Sustainable Development Report
67. Phuoc Minh N, Thanh Vo T, Van Man L, Dinh Phong T, Van Toan N, Hoang Nam V (2019) Green pigment extraction from Pandan (Pandanus Amaryllifolius) and its application in food industry
68. Picchio R, Latterini F, Venanzi R, Stefanoni W, Suardi A, Tocci D, Pari L (2020) Pellet production from woody and non-woody feedstocks: a review on biomass quality evaluation. Energies (Basel) 13
69. Pode R (2016) Potential applications of rice husk ash waste from rice husk biomass power plant. Renew Sustain Energy Rev 53:1468–1485
70. RAIS (2023) RAIS 2022—Ministério do Trabalho e Emprego. https://www.gov.br/trabalho-e-emprego/pt-br/assuntos/estatisticas-trabalho/rais/rais-2022. Accessed 5 Oct 2024
71. Rocha DHD, Sakamoto IK, Varesche MBA (2023) Optimization of fermentative parameters to improve hydrogen production: is the co-fermentation of waste from the citrus agro-industrial an interesting alternative for energy recovery? J Environ Chem Eng 11(6). https://doi.org/10.1016/j.jece.2023.111252
72. Rodrigues DB, Flores ÉMM, Barin JS, Mercadante AZ, Jacob-Lopes E, Zepka LQ (2014) Production of carotenoids from microalgae cultivated using agro-industrial wastes. Food Res Int 65(PB):144–148. https://doi.org/10.1016/j.foodres.2014.06.037
73. Ruml A, Chrisendo D, Iddrisu AM, Karakara AA, Nuryartono N, Osabuohien E, Lay J (2022) Smallholders in agro-industrial production: lessons for rural development from a comparative analysis of Ghana's and Indonesia's oil palm sectors. Land Use Policy 119:106196. https://doi.org/10.1016/J.LANDUSEPOL.2022.106196
74. Sadiq M, Adil M, Paul J (2021) An innovation resistance theory perspective on purchase of eco-friendly cosmetics. J Retail Consum Serv 59:102369. https://doi.org/10.1016/J.JRETCONSER.2020.102369
75. Salvador R, Barros MV, Donner M, Brito P, Halog A, De Francisco AC (2022) How to advance regional circular bioeconomy systems? Identifying barriers, challenges, drivers, and opportunities. Sustain Prod Consum 32:248–269. https://doi.org/10.1016/J.SPC.2022.04.025
76. Sang Y, Yu H, Han E, Suh M-J, Bae S, Sang Y, Yu H, Han E (2022) Understanding the barriers to consumer purchasing of zero-waste products. Sustainability 14(24):16858. https://doi.org/10.3390/SU142416858
77. Sangmesh B, Patil N, Jaiswal KK, Gowrishankar TP, Selvakumar KK, Jyothi MS, Jyothilakshmi R, Kumar S (2023) Development of sustainable alternative materials for the construction of green buildings using agricultural residues: a review. Constr Build Mater 368:130457. https://doi.org/10.1016/J.CONBUILDMAT.2023.130457
78. Santos EO, Costa CC, Gonzáles PLS, Ribas PP, Batista CE (2024) Agro-industrial waste of guarana as potential coagulant agent for wastewater treatment. J Water Process Eng 67. https://doi.org/10.1016/j.jwpe.2024.106165
79. Sarangi PK, Srivastava RK, Vivekanand V, Goksen G, Sahoo UK, Thakur TK, Debeaufort F, Uysal-Unalan I, Pugazhendhi A (2024) Recovery of green phenolic compounds from lignin-based source: role of ferulic acid esterase towards waste valorization and bioeconomic perspectives. Environ Res 256:119218. https://doi.org/10.1016/J.ENVRES.2024.119218
80. Schiebel CS, Bueno LR, Pargas RB, de Mello Braga LLV, da Silva KS, Fernandes ACVU, dos Santos Maia MH, de Oliveira NMT, Bach C, Maria-Ferreira D (2024) Exploring the biological activities and potential therapeutic applications of agro-industrial waste products through non-clinical studies: a systematic review. Sci Total Environ 950:175317. https://doi.org/10.1016/J.SCITOTENV.2024.175317
81. Senado Federal Brasileiro (2024) Projeto de Lei n° 1874, de 2022. https://www25.senado.leg.br/web/atividade/materias/-/materia/153918. Accessed 29 Oct 2024

82. Shah AV, Singh A, Sabyasachi Mohanty S, Kumar Srivastava V, Varjani S (2022) Organic solid waste: biorefinery approach as a sustainable strategy in circular bioeconomy. Bioresour Technol 349:126835. https://doi.org/10.1016/J.BIORTECH.2022.126835
83. Sharma S, Basu S, Shetti NP, Kamali M, Walvekar P, Aminabhavi TM (2020) Waste-to-energy nexus: a sustainable development. Environ Pollut 267:115501. https://doi.org/10.1016/J.ENVPOL.2020.115501
84. Shin SG, Han G, Lim J, Lee C, Hwang S (2010) A comprehensive microbial insight into two-stage anaerobic digestion of food waste-recycling wastewater. Water Res 44(17):4838–4849. https://doi.org/10.1016/j.watres.2010.07.019
85. Shinde R, Shahi DK, Mahapatra P, Singh CS, Naik SK, Thombare N, Singh AK (2022) Management of crop residues with special reference to the on-farm utilization methods: a review. Ind Crops Prod 181:114772. https://doi.org/10.1016/J.INDCROP.2022.114772
86. Silva JBMD, Paiva MTP, Pavanello ACL, Mantovan J, Mali S (2022) Fiber-rich ingredients obtained from agro-industrial residues through combined hydrothermal-chemical processes. Food Chem Adv 1. https://doi.org/10.1016/j.focha.2022.100149
87. Sri Shalini S, Palanivelu K, Ramachandran A, Raghavan V (2020) Biochar from biomass waste as a renewable carbon material for climate change mitigation in reducing greenhouse gas emissions—a review. Biomass Convers Biorefinery 11(5):2247–2267. https://doi.org/10.1007/S13399-020-00604-5
88. Tariq A, Bhawani SA, Moheman A, Alotaibi KM (2023) Introduction to agro-industrial waste. In: Extraction of natural products from agro-industrial wastes: a green and sustainable approach. Elsevier, pp 1–18
89. Tirkolaee EB, Mahdavi I, Esfahani MMS, Weber GW (2020) A robust green location-allocation-inventory problem to design an urban waste management system under uncertainty. Waste Manage 102:340–350. https://doi.org/10.1016/J.WASMAN.2019.10.038
90. Tomé AG, Ribeiro EAM, Lima M, Brocenschi RF, Ribeiro LNM, Amaral e. FA (2023) Biorefinery of peanut shell agro-industrial lignocellulosic waste for synthesis of a natural coagulant applied in the treatment of dairy wastewater. J Environ Chem Eng 11(6). https://doi.org/10.1016/j.jece.2023.111535
91. Ucella-Filho JGM, Ferreira NS, Cupertino GFM, Guedes NA, de Paula H, Kitagawa RR, Balu AM, Luque R, Ignacchiti MDC, Parreira LA, Resende JA, Dias Junior AF (2024) Antiviral potential of bio-oil from Citrus sinensis waste wood as a therapeutic approach against COVID-19. J Clean Prod 447:141583. https://doi.org/10.1016/J.JCLEPRO.2024.141583
92. United Nations Development Programme (2024) Objetivos de Desenvolvimento Sustentável. https://www.undp.org/pt/brazil/objetivos-de-desenvolvimento-sustentavel
93. United Nations Environment Programme (2024) Food Waste Index Report 2024. Nairobi
94. United Nations Environment Programme (2024) Global waste management outlook 2024—beyond an age of waste: turning rubbish into a resource. United Nations Environment Programme, Nairobi
95. Uzoagba CEJ, Okoroigwe E, Kadivar M, Anye VC, Bello A, Ezealigo U, Odette Ngasoh F, Pereira H, Azikiwe Onwualu P (2024) Characterization of wood, leaves, barks, and pod wastes from Prosopis africana biomass for biofuel production. Waste Manage Bull 2(3):172–182. https://doi.org/10.1016/j.wmb.2024.07.007
96. Van de Velden M, Baeyens J, Brems A, Janssens B, Dewil R (2010) Fundamentals, kinetics and endothermicity of the biomass pyrolysis reaction. Renew Energy 35(1):232–242. https://doi.org/10.1016/j.renene.2009.04.019
97. Van Wayenbergh E, Struyf N, Rezaei MN, Sagalowicz L, Bel-Rhlid R, Moccand C, Courtin CM (2020) Cereal bran protects vitamin A from degradation during simmering and storage. Food Chem 331. https://doi.org/10.1016/j.foodchem.2020.127292
98. Wagh MS, Sowjanya S, Nath PC, Chakraborty A, Amrit R, Mishra B, Mishra AK, Mohanta YK (2024) Valorisation of agro-industrial wastes: circular bioeconomy and biorefinery process—a sustainable symphony. Process Saf Environ Prot 183:708–725. https://doi.org/10.1016/J.PSEP.2024.01.055

99. Wani KA, Mamta RRJ (2013) Bioconversion of garden waste, kitchen waste and cow dung into value-added products using earthworm Eisenia fetida. Saudi J Biol Sci 20(2):149–154. https://doi.org/10.1016/j.sjbs.2013.01.001
100. Weilack I, Mehren L, Schieber A, Weber F (2023) Grape-derived pectic polysaccharides alter the tannin and pigment composition of Cabernet Sauvignon red wines. Curr Res Food Sci 6. https://doi.org/10.1016/j.crfs.2023.100506
101. Xu C, Asraful Alam M, Wang Z, Chen H, Zhang J, Huang S, Zhuang W, Xu J (2021) Mechanisms of bio-additives on boosting enzymatic hydrolysis of lignocellulosic biomass. Bioresour Technol 337:125341. https://doi.org/10.1016/J.BIORTECH.2021.125341
102. Yaqoob AA, Ibrahim MNM, Alshammari MB, Ahmad A, Umar K, Rashid M (2023) Extraction of lignin from agro-industrial waste. In: Extraction of natural products from agro-industrial wastes: a green and sustainable approach. Elsevier, pp 217–232
103. Zhang QW, Lin LG, Ye WC (2018) Técnicas para extração e isolamento de produtos naturais: uma revisão abrangente. Chin Med 13:20
104. Zang Z, Tang S, Li Z, Chou S, Shu C, Chen Y, Chen W, Yang S, Yang Y, Tian J, Li B (2022) An updated review on the stability of anthocyanins regarding the interaction with food proteins and polysaccharides. Compr Rev Food Sci Food Saf 21(5):4378–4401. https://doi.org/10.1111/1541-4337.13026